Agriculture and Environment for Developing Countries

Published by

Mzuni Press
P/Bag 201
Luwinga, Mzuzu 2
Malawi

ISBN 978-99960-27-09-3 (Mzuni Press)

Mzuni Press is represented outside Africa by:
African Books Collective Oxford (also for e-books)
(orders@africanbookscollective.com)

www.mzunipress.luviri.net
www.africanbookscollective.com

Printed in Malawi by Baptist Publications, P.O. Box 249, Lilongwe

Agriculture and Environment for Developing Countries

B.H.Z. Moyo

Mzuni Press

Mzuni Books no. 10

2014

Content

Chapter 1

SUSTAINABLE AGRICULTURE

1.0 Introduction

Sustainability implies persistence and the capacity of something to continue for a long time. In the case of agriculture, it also implies not damaging or degrading natural resources and the environment upon which it (agriculture) depends. Agricultural activities have always been based on sustaining production for the benefit of mankind for as long as human-beings exist.[1] For example, shifting cultivation practices involve moving cultivation sites in order to rest those (sites) that have become tired with the purpose of sustaining production. Furthermore, farmers who graze animals in the bush give chance for pastures to regenerate through rotational grazing. Sustainability, therefore, means keeping the community alive, that is achieved through meeting the requirements of producing sufficient agricultural food with greater efficiency.[2] This creates the importance of ecologically acceptable production that is realized when everything removed is then replaced so as not to harm the ecological system.[3] In addition, sustainability also means a thriving economic and social order with production structures and relationships which ensure a fair distribution of income, power and opportunities, thus providing the basis for social peace. Indeed, sustainability of agricultural production has been a necessary goal for time in memorial particularly under indigenous means of production.[4]

---

[1]See C. Eriksen, "Why do they Burn the 'Bush'? Fire, Rural Livelihoods, and Conservation in Zambia," *The Geographical Journal* 173 (3), 2007, pp. 242-256, 2007; J.A. Riseth, "An Indigenous Perspective on National Parks and Sami Reindeer Management in Norway," *Geographical Research* 45 (2), 2007, pp. 177-185.

[2]C. Lado, "Sustainable Environmental Resource Utilisation: a Case of Farmers' Ethno Botanical Knowledge and Rural Change in Bungoma District, Kenya," *Applied Geography Journal* 24, 2004, pp. 281-302.

[3]Ibid.

[4]Ibid.

The important point is to clarify what is being sustained, for how long, for whose benefit, at whose cost and at what cost, over what area and measured by what criteria? Answering the above questions can be difficult as it may mean assessing and trading of values and benefits. For example, the grazing of animals in the bush by pastoralists and shifting cultivation have generated debate among scholars that relate to differing values and norms between those who engage in these two activities and those who try to assess, examine and evaluate such activities. Pastoralism and shifting cultivation practices have been classified by some agricultural experts as activities that contribute to the destruction of natural resources and degradation of the environment.[5] However, it has been demonstrated that shifting cultivation helps in restoring nutrients depleted as a result of cultivation.[6] Natural processes of nutrient restoration from decaying plants and regeneration of species occur (see Figure 1.1)

The scientific understanding of the importance of pastoral mobility has changed over time. The basis for studying dry land ecosystem had been Clement's model of vegetation succession, and the idea of equilibrium ecosystems and livestock density dependent limitations of primary production.[7] Pastoral development studies have predominantly operated

---

[5] See J. Briggs, M. Badri, and A.M. Mekki, "Indigenous Knowledges and Vegetation Use among Bedouins in the Eastern Desert of Egypt." *Applied Geography* 19, 1999, pp. 78-103, B.H.Z. Moyo, "The Use and Role of Indigenous Knowledge in Small-scale Agricultural Systems in Africa: the Case of Farmers in Northern Malawi," Published PhD thesis. Glasgow University, 2008. http://theses. gla.ac.uk and S. Kratli, "What do Breeders Breed? On Pastoralists, Cattle and Unpredictability", *Journal of Agriculture and Environment for International Development* 102 (1-2), 2008, pp. 123-139.

[6] B.H.Z. Moyo, "The Use and Role of Indigenous Knowledge in Small-scale Agricultural Systems in Africa: the Case of Farmers in Northern Malawi," Published PhD thesis. Glasgow University, 2008. http://theses.gla.ac.uk

[7] See H.K. Adriansen, "Understanding Pastoral Mobility: the Case of Senegalese Fulani," *The Geographical Journal* 174 (3), 2008, pp. 207-222.

Figure 1.1 A fallow for more than 10 years

from a range management perspective that puts importance on the carrying capacity of the land. Pastoralists are thus seen to damage the environment through overstocking that is beyond the carrying capacity of the pastures on the land. An equilibrium concept, such as the carrying capacity, is not relevant in disequilibrium system, since livestock numbers seldom reach densities high enough to influence vegetation productivity Clearly, there is a difference between some experts and practitioners of pastoralism and shifting cultivation in terms of the basic understanding of the environment and the consequent uses, value and norms.[8]

It is important that sustainable agriculture does not prescribe a concretely defined set of technologies, practices or policies while being fully aware that as conditions and knowledge change development experts, farmers and communities adapt and those that may not should be encouraged and allowed to change and adapt.[9]

---

[8]See Ibid.

The basic challenge for sustainable agriculture is to utilize the available natural and human resources. This can be achieved through the intensive and extensive use of internal inputs while reducing external inputs.[10] Internal inputs can be regenerated to ensure their efficient and effective use in order to attain sustainability of agricultural production. In sustainable agriculture the use of available resources and the reduction or even in some instances the elimination of dependency on external systems (including inputs) is an important factor to consider.

1.1 Sustainable Agricultural Production

Sustainable agricultural production has mainly referred to increased production and productivity to feed the growing global population and also to maximize profit.[11] Green revolution became a term to refer to this process in the later years between 1960 and 1970. Although in the past it was farmers who strived to increase agricultural production, the introduction of 'formal' science (in agriculture) led to the process of increasing agricultural production becoming a top down approach involving development experts providing technologies to farmers for free or sale that are suitable for increasing agricultural output (both through production and increased productivity). Indeed, Brady notes that the farmer is practical,

---

[9]J. Briggs, J. Sharp, H. Yacoub, N. Hamed and A. Roe, "Environmental Knowledge Production: Evidence from the Bedouin Communities in Southern Egypt." *Egypt Journal of International Development* 19, 2007, pp. 239-251; B.H.Z. Moyo, "The Use and Role of Indigenous Knowledge in Small-scale Agricultural Systems in Africa: the Case of Farmers in Northern Malawi." Published PhD thesis. Glasgow University,, 2008. http://theses.gla.ac.uk; B.H.Z. Moyo, "Indigenous Knowledge-Based Farming Practices: A Setting for the Contestation of Modernity, Development and Progress," *Scottish Geographical Journal* 125 (3-4), 2009, pp. 353-360; S. Kratli, "What do Breeders Breed? On Pastoralists, Cattle and Unpredictability," *Journal of Agriculture and Environment for International Development* 102 (1-2), 2008, pp. 123-139.

[10]S. Gomes de Almeida, and G.B. Fernandes, "Economic Benefits of a Transition to Ecological Agriculture. Changing Farming practices," *LEISA Magazine* 22 (2), 2006, pp. 28-29.

[11]See N.C. Brady, "The Nature and Properties of Soils," New Jersey, Prentice Hall, 1974, p. 3.

having the production of food and fibre as an ultimate goal but at the same time, he/she must be a scientist to determine reasons for variation in the productivity of soils and to find means of conserving and improving this productivity.[12] Some of the guiding philosophies in increasing agricultural production are economies of scale, the limiting factors of production and the principle of comparative advantage. Such ideologies led to increased areas under cultivation, specialization of production in areas most suitable and production of commodities best suited to such areas, mono-cropping, intensive use of external inputs such as inorganic fertilizers, pesticides (insecticides, fungicides, herbicides, rodenticides and nematocides) and high yielding varieties of crops and high yielding breeds of animals; and intensive use of fossil fuels to provide power. However, it has become clear that such type of production is damaging to the environment. Brady demonstrates that the death of creatures such as birds and fish from the accumulation of pesticides in their bodies sound the warning cry that man must know more about the ecological effects and the danger posed by pesticides on the environment.[13] Furthermore, there is mounting evidence that increased use of inorganic fertilizers and pesticides does not sustain agricultural production. Indeed, it is clear, for example, that India's agriculture is now experiencing problems with soils responding only grudgingly to the heavy use of fertilizers after their successful use in the Green Revolution.[14] In some instances there is need for increasing amounts of inorganic fertilizers and pesticides just to maintain production at constant levels. Furthermore, the increased yields of crops mean an increase in the transfer of minerals from the soils in form of yields to areas of consumption requiring the additional sources of nutrients to replenish those harvested in form of commodities.[15] Additionally, the growing of uniform crops (one crop and one variety) over large areas result in reduced diversity. The reduced diversity increase the scale and intensity of pests and diseases in crops and livestock especially when the resistance to pests and

---

[12]Ibid, p. 6.

[13]Ibid, p. 553.

[14]D.D. Phiri, "D.D. Phiri Column, Lessons from Green Revolution," *The Nation Newspaper* of 26th March, 2010.

[15]See N.C. Brady, "The Nature and Properties of Soils," New Jersey, Prentice Hall, 1974, pp. 4, 20; N. Hudson, *Soil and Water Conservation*. Bedford, Silsoe Associates Ampthill, 1981, p. 22.

diseases decrease over time either by natural phenomenon or as a result of pests and diseases mutating and becoming more virulent. Some pests develop resistance to chemicals necessitating higher dosages or the development of new chemicals to replace those to which the pests are resistant.[16]

1.2 Characteristics of Sustainable Agriculture

Sustainable agriculture thus becomes a system of production (food, fibre, commodities) that takes advantage of the following:

- Incorporation of natural processes such as nutrient recycling as in the case of incorporation of crop residues and animal wastes in cultivated areas, nitrogen fixation, escaping and avoiding pest and disease to improve agricultural production processes that are 'profitable[17]' and efficient.[18]

- A reduction in the use of those external and non-renewable inputs with greatest potential to damage the environment, harm the health of farmers as well as consumers such as pesticides and inorganic fertilizers.

- The recognition and use of farmers and rural people in problem analysis and technology development, adaptation and dissemination (extension).

- Equitable access to productive resources such as land and opportunities so that there is attainment of socially just forms of agriculture.

- Productive use of local knowledge and practices, including innovative approaches generated by local practitioners (farmers). For example, Moyo and Kratli independently found out that farmers'

---

[16]N.C. Brady, "The Nature and Properties of Soils," New Jersey, Prentice Hall, 1974, p. 553.

[17]Profitability here refers to production that meets the requirements of the household as well as actual profit making.

[18]See B.H.Z. Moyo, "The Use and Role of Indigenous Knowledge in Small-scale Agricultural Systems in Africa: the Case of Farmers in Northern Malawi," Published PhD thesis. Glasgow University, 2008. http://theses.gla.ac.uk

production strategy is aimed at exploiting the variability of an unpredictable environment.[19] Farmers' production systems do not necessarily try to conquer nature but to work within its limits as opposed to the basis for scientific production methodologies that try to conquer nature as manifested by the production of genetically modified crops and livestock (biotechnology). Clearly farmers are not risk averse but prudent risk managers.

• Restoration of self-reliance amongst farmers and rural people. This is an old aim of small-scale farmers. For example, Buchanan (1885) reports of adequate agricultural production in Nyasaland without the use of external inputs. In recent times it has been demonstrated that poverty becomes a reality only when the self-reliance and cultural practices of sharing resources of Third World Countries are replaced with modernization.[20]

When most of these characteristics are fulfilled, the resultant agricultural production then becomes ecologically sound, economically viable, socially just (without exploitation of the proletariats), humane (respecting the lives of plants and animals, incorporation and acceptance of human values such as trust, honesty, self respect, cooperation and being compassionate) and adaptable. There are examples of sustainable agricultural practices on the African continent today. For instance, the WoDaaBe in Niger using local knowledge organize their cattle into matrilineal lineages and operate selection within (but not between) lineages so that they nurture and structure diversity within the breeding population.[21]Kratli, (2008) notes that such type of selection ensures that all lineages are up to the task of performing well within the household's production strategy, at each point in time each lineage embedding a specific configuration of animal-human-environment interaction.[22] In pursuit of sustainability and in particular of

---

[19]Ibid; S. Kratli, "What do Breeders Breed? On Pastoralists, Cattle and Unpredictability," *Journal of Agriculture and Environment for International Development* 102 (1-2), 2008, pp. 123-139.

[20]Arturo Escobar, *Encountering Development: the Making and Unmaking of the Third World*, Princeton University Press, 1995.

[21]S. Kratli, "What do Breeders Breed? On Pastoralists, Cattle and Unpredictability," *Journal of Agriculture and Environment for International Development* 102 (1-2), 2008, pp. 123-139.

[22]Ibid.

being humane the WoDaaBe exploit both genetic and extra-genetic inheritance in their cattle breeding population. They use the animals' capacity to actively engage with their environment (including conspecifics and humans), generating responsive change during their lifetime, and transmitting such resources along social networks. The pastoralists generate lineage duration, which is when a lineage gives consistently good performance within the production strategy over an extended period of time including events of severe stress as found in dry lands of Africa. Similarly, in Malawi farmers use local knowledge to ensure food security by planting crops over an extended period of up to three months (November to January) within a given rainy season so that the varying times enable farmers stabilize total production at the end of each year which would otherwise fluctuate so much if planting was only done with the first planting rains (especially when dry spells occur).[23] In addition, the varying of planting time of crops enables farmers to escape pests and diseases. Late planting in Mzuzu (a place in northern part of Malawi) ensures avoidance of maize stalk-borer infestation, for example. In addition, farmers in Mzuzu use the variations in weather conditions and soil types to sustain yields over the course of cultivation throughout their life.[24] There are thus embedded climate and weather adaption practices within farmers' local knowledges.

Sustainable agriculture is a necessity for mankind to live in harmony with Mother Nature while continuing to benefit from it with minimum negative impacts on himself and the general environment. It becomes apparent that indigenous agricultural systems demonstrate a considerable knowledge of, and sympathy with, the environment. However, it must be clarified here that the diversity of mankind's behaviour and the consequent agricultural practices may in some cases lead to local knowledge that leads to environmental degradation. The trick is to research on and promote those indigenous agricultural practices that lead to the protection and conservation of resources and the environment while remaining highly productive and maintaining diversity.

---

[23]B.H.Z. Moyo, "The Use and Role of Indigenous Knowledge in Small-scale Agricultural Systems in Africa: the Case of Farmers in Northern Malawi," Published PhD thesis. Glasgow University, Glasgow UK, 2008. http://theses.gla.ac.uk.

[24]See Ibid.

1.3 The Role of Indigenous Knowledge in Sustainable Agriculture: Experiences from Agricultural Practices in Northern Malawi

This section looks at agricultural practices in the northern Malawi which show how farmers strive to maintain and improve production while conserving resources and protecting and managing the environment. The case study is based on indigenous knowledge associated with sustainable agricultural practices in Northern Malawi while noting what other scholars have found elsewhere on the same. For example, some of the literature would suggest that indigenous knowledge is developed and re-worked to suit the specific environment, needs and priorities of the farmers.[25] This case study, therefore, critically examines various environments in which farmers cultivate crops. Their description of garden types and the reasons behind their nomenclature is analyzed. The practices farmers use in cultivating such sites and their knowledge about soils and soil fertility is examined and compared with scientifically proven knowledges and linked to sustainability. Farmers' efforts to understand and manage soils are examined in detail, based on the use of inorganic fertilizers and organic matter from crop residues and natural vegetation. The role culture plays in indigenous knowledge development and the corresponding farmers' understanding of their environment is contextualized in relation to sustainable agriculture. Contextualization of indigenous knowledge helps in understanding farmers' decision-making rationalities.

1.3.1 Garden Types

All smallholder farmers included in this study have three types of gardens, which is a common feature in the study area. It is an accepted tradition that a household should have a *dimba,* which is a garden in a nearby wetland;

---

[25]C. Beckford, D. Barker and S. Bailey, "Adaptation, Innovation and Domestic Food Production in Jamaica: Some Examples of Survival Strategies of Small Scale Farmers." *Singapore Journal of Tropical Geography* 28, 2007, pp 273-286; J. Briggs, J. Sharp, H. Yacoub, N. Hamed and A. Roe, "Environmental Knowledge Production: Evidence from the Bedouin Communities in Southern Egypt," *Egypt Journal of International Development* 19, 2007, pp. 239-251.

a*khonde*, which is a garden close to the house; and a *mundaukulu,* which is a main garden where the bulk of food production is undertaken. The *mundaukulu* is normally located at distances from dwelling houses that range from as little as 100 metres to 20 km or further, mainly in areas considered to be very fertile. Many streams in the area make it possible for almost all farmers to have at least one *dimba* each; the village settings allow for each house to have a *khonde* garden, and a *mundaukulu* is a necessity in a situation where almost all food is produced and consumed by the grower. *Khonde* gardens are purposefully created by leaving spaces for cultivation between the houses in villages. Effort is made by farmers, therefore, to have these three types of gardens. Each name has a special meaning that is a description of the garden, relates to its geographical features, or refers to the purpose for which the crops are grown in such fields. The local common name for all types of gardens is *munda*, which denotes the main garden when the adjective *'ukulu'* is added to it, and therefore is called *mundaukulu* which means a big garden. Virgin land which is converted to farm land in its first year is called *nthewele* or *mphangula.*

A garden that is along a stream or river bank, which is inundated in the rainy season, is called a *dimba*. Such gardens are prepared in April to August and cultivated in the dry season from June to September, although some farmers extend this period by cultivating *dimba* from May up to the end of the dry season, which is late in the month of November. This is when the water table is low and the soil is not water-logged. Here is where the last annual crop is cultivated, such as maize, which is harvested in December just before inundation recurs as a result of the rains. Crops grown in these gardens rely on moisture residues and may be supplemented with water from the stream, using tools such as watering cans. A garden along a river bank that has good drainage and is cultivated in the rainy season is called *njota*. The major difference between *dimba* and *njota* is the time when crops are planted, which is during the rainy season in a *njota* and the dry season in a *dimba*. This is influenced and controlled by the soil moisture regime rather than soil fertility. The *dimba* is inundated in the rainy season and thus unsuitable for crops that need well-aerated soils, such as maize, while the *njota* remains free of inundation. Hence, its soils are well-aerated and remain suitable for crop production, especially maize, in a similar manner to upland gardens that are cultivated in the rainy season. These field types are also important in terms of the crops grown, their use and the cultivation practices followed by farmers. For example,

14

njota are considered in the same way as *mundaukulu*, and, as such, food grown here, particularly maize, is mainly stored and used throughout the year, whereas maize grown in *dimba* is eaten as green maize soon after its maturity or is used soon after harvest in December, when it is dry enough to be processed into flour. Farmers' names given to field types are at times influenced by crop use in addition to names given based on location. The diverse environmental characteristics of cultivation sites play a major role in the names farmers give to their fields.

The classification of field types is also based on the physical position of the garden, as well as its size, making it complicated to understand to those unaware of the many factors farmers take into account in giving names to their fields. Indeed, when the area of the garden is quarter of a hectare or less, and is next to a dwelling house, it retains the name of *khonde*. However, a *khonde* garden that is bigger than half a hectare is referred to as a main garden, despite its location. It is important to note that despite this being referred to as a main garden, it is a necessary requirement for most farmers to have a main garden away from the *khonde* area for food security reasons. Farmers believe that a garden away from the houses has the advantage that crops grown here are less likely to be easily accessed because of the distance, and therefore are only used when they mature, become dry, are harvested and then put in storage. The distance between the main garden and the village then ensures that households are food secure for the whole year because food produced here is used only after it has been put in storage, which is when it is dry enough to store well up to the next harvest. For most seed crops, such as maize, this is when its moisture content is 14% of its total dry matter weight. The farmers achieve this moisture content not by drying and weighing the grains, as is the norm in Western scientific methods, but by the length of drying in the sun, then using touch and feel combining with the sounds grains make when shaken or twisting the cob. For farmers, dry grains make a special sound 'zikutiwayawaya' that is distinct from those that are not fully dry. This sound is similar to stones shaken in a container. Farmers make the allusion that stones make such a sound when they are dry, so that grains making a similar 'noise' mean they are dry enough for storage.

Crops grown in *khonde* gardens are meant to supplement what is produced in the main gardens, and they are used as soon as they mature. For example, mature green maize is picked and eaten roasted or boiled in water immediately, with very little left over as dry maize suitable for use

15

and storage in the dry season. Children are also allocated land to practice cultivation in *khonde* gardens, almost as a training ground for children to become productive farmers in the future. This is where they develop skills and practices such as land clearing, ridging, planting, weeding and harvesting necessary for farming in preparation for their adult lives. Children are allowed to cultivate whatever crops they choose on the sections of *khonde* allocated to them. *Khonde* gardens, therefore, have the highest total number of individual crops (14) grown by farmers as shown in Table 1.0, compared to main gardens (10) and *dimba*(9), a reflection of the 'experimentation', teaching and learning processes being carried out in these locations in addition to the provision of a wide variety of (perishable) food types.

Table 1.0 Information on types of gardens smallholder farmers cultivate and number of crops grown in each garden

Garden type	Mean number of crops per garden type	Total number of different crops grown by farmers in the study area	Percentage of total crops grown
Mundaukul u	10	19	53
Khonde	14	19	74
Dimba	9	19	47
N=111			

The re-working and production of indigenous knowledge is continuous and involves 'experimentation' and making observations. The *khonde* gardens are intensively used for knowledge production because conducting 'experiments' and making observations is easy due to their closeness to dwelling houses. For example, one smallholder farmer dry-planted four maize seeds in November 2006 in his *khonde* garden which grew to be 30 cm taller at maturity than the maize planted a month later in the same field; they also averaged two to three maize cobs per plant, all of which were bigger than those planted a month later with the first rains in the same garden. The later-planted maize had only one cob per plant. This is very important to the farmers, as it demonstrates the yield potential of dry-planted maize. It was not surprising to hear from this particular farmer that a bigger area around the *khonde* was grown with dry-planted maize in the following season. I was shown this crop when a third visit was made in February 2007. The rest of the maize crops in the *khonde* were smaller in size and height than those that were dry-planted, a reflection of the importance of early planting in this area that takes advantage of soil nutrients before they are leached down beyond a crop's roots by rain water later in the season.

The major problem with indigenous knowledge is that it is 'rarely' published and advertised by the producers. For example, the farmer conducting this 'trial' only alerted me about his 'experiment' on my third visit, and perhaps then only because it was a success. It is, therefore, safe to assume that many crop and cultivation 'experiments' conducted by farmers may go unnoticed and are not exposed to outsiders, especially if and when they are considered unsuccessful.

Main gardens (*mundaukulu*) are located away from villages in most cases. They comprise a larger area in size, ranging from 1 hectare to 20 hectares with a mean size of 3.9 hectares as compared to *dimba* and *khonde,* which are normally less than a half a hectare. Up to ten food crops are grown in the main gardens under a mixed cropping pattern dominated by maize, but mixed with stands of beans, pumpkins, bananas, vegetables, cassava and mangoes, for example, making them a complex 'experimental' site for knowledge production. Very few farmers (only three) grow tobacco in their main gardens, despite it being a major cash and export crop in Malawi.

It is common practice for farmers to own more than one *mundaukulu*, as shown by the mode being 3 gardens and the mean being 2.9 (Table 1.1).

Table 1.1 Information on number of main gardens smallholder farmers own and/or cultivate

Number of main gardens	Number of farmers	Percentage
1	7	6.3
2	31	27.9
3	48	43.2
4	15	13.5

5	9	8.1
6	1	0.9
Mean number of gardens 2.92		
Median number of gardens 3.00		
Mode *3*		

There are several reasons for farmers to have more than one main garden, including ensuring food security by having gardens in different areas of different soil types and microclimates; ensuring enough land is owned that will eventually be passed on to children; and the cultivation of land inherited from parents. In cases where gardens have been inherited from parents, the current farmers have not been able to select the sites according to their own preferences and soil fertility criteria. The act of choosing the site was done for them by parents, or even earlier generations, but it is reasonable to assume that parents based the choice of such gardens on soil fertility indicators that are now utilized by their children in selecting new garden sites. Such sites in different locations render themselves ideal for observation and comparison of crop performances both within one season and across many seasons that the crops have been grown, based on location, niche and local climate. These different gardens resemble trials conducted by researchers after a seed variety has been developed at a research station and is then grown on a wider scale in various parts of the country under a technical name called 'variety trials'. Variety trials are used to ascertain crop performances in the field, after proving either to be high yielding or resistant to diseases, depending on the objective of the research carried out on research stations. It is difficult to ignore some similarities between traditional knowledge and Western technologies' development and production procedures and processes.

The study area has predominantly red soils (latosols) which are considered to be relatively infertile because they have a low nitrogen content, a point also confirmed from the analysis I conducted in the 2006/2007 cropping season. There are, however, patches of soils that are considered to be more fertile by farmers using their own fertility indicators such as soil colour and vegetation, and these are sought and opened up as farmlands. Furthermore, farmers recognize that under their customs, land is passed on to children through inheritance of pieces of land cultivated by parents. Effort is made to open virgin land to establish ownership by cultivation for the purpose of subsequently passing that land to children. The more children a farmer has, the more land is opened up for passing to the children of the household later in life, particularly if they are male. Female children are expected to marry and inherit land through their husbands.

Those food crops, particularly maize, which are grown in *mundaukulu*, are rarely eaten when mature as green maize, while maize grown in *khonde* is eaten as green maize as soon as it is mature. Maize in *mundaukulu* is left to dry and is then harvested when the grain moisture content is suitable for storage, which is usually below 20% of individual grain weight or about 14% of its total dry matter. Farmers said that when maize is dry enough, it is lighter and easy to carry to granaries, and can be stored well up to the next harvest without mould infestation. To achieve this desirable moisture content, farmers cut the dry whole maize plant and create several circular upright heaps called maize bundles that stand on the stem base in gardens (Figure 1.1). Such maize heaps are known as *mukukwe*, and typically have a diameter of three to four metres. The maize cobs are removed by hand from the plants during the harvesting process when the cobs are considered to be sufficiently dry. This moisture content is usually achieved after the maize has been placed in the heap for about one month or more. The conical heaps are designed to allow air to pass through, collecting moisture which the heat from the sun then removes from the grains in the cob. The approximately uniform size of heaps in all gardens, therefore, is a reflection of the optimum quantity of maize plants that can be effectively dried under such conditions. A similar practice was common in Scotland in the early 20th century, where grain was dried in the open while standing on cut plant stalks placed in conical heaps.[26] It is important to note that this is a practice associated with maize only in the main gardens. Maize in *khonde* gardens is harvested without making conical heaps and is allowed to dry while still in planting stations, reflecting the fact that it is consumed soon after harvest, and therefore it does not need to be thoroughly dried for storage.

---

[26]J. Black, *'O ye Green Memories O' The Auld Days. 'A Renfrewshire Farmer's Story.* Dundee: Accolade Publishing, 2006.

Figure 1.2 Maize being dried in conical shaped heaps before being harvested

Maize in the main gardens is rarely harvested whilst still green and this is achieved by the very nature of their location. Picking maize cobs while green in *khonde* is an activity that is largely done for pleasure because green maize is consumed as a snack so that it is often seen as a leisure activity that is particularly done, in most cases during morning or late afternoon hours. The long distances between the main gardens and the villages, therefore, require that effort is made to access maize in such gardens, and, as such, the leisure aspect is lost. There is a second aspect of *khonde* that is not related to green maize harvesting, however, but it is related to work output. Once the farmers have left the village for work, there is a reduced chance of being called to attend to minor events or to greet visitors, for example, and they can hence concentrate on their farming activities in the *mundaukulu*. For example, one farmer shrewdly said that producing enough food for the family requires avoiding social events during critical cultivation times, and this is achieved by being away working in *mundaukulu* that are located far from the dwelling houses. Indigenous knowledge is sophisticated in nature and is craftily and wisely used to ensure food security at household level as shown by the sometimes deliberate creation of considerable distances to *mundaukulu.*

Maize grown in the *mundaukulu* is reserved for use as flour upon drying and processing, and is then eaten cooked as a thick porridge called *sima.* Every effort is made to achieve household food security and the distance between the village and the *mundaukulu* plays a positive role in this respect. The fact that the *mundaukulu* are far from the village makes it possible to work without disturbance and this results in an improved work output from the labour input in activities like land preparation, ridging and weeding. Consequently, the yields of crops increase, mainly as a result of the timely completion of farming operations. The increase in output as a result of these timely operations is then enhanced through the avoidance of harvesting green maize, as happens in the *khonde,* which would reduce the harvest of dry matured maize for storage and consumption for the rest of the year after the harvest. It can be argued that food sufficiency is achieved through a deliberate strategy using the distance barrier between dwelling houses and *mundaukulu.* The food self-sufficiency ensured by these considerable distances between dwelling houses and *mundaukulu* contrasts with the Western ways of knowing that advocates the achievement of timely farming operations by 'amalgamating' of scattered gardens in order to reduce time for travelling between them.

Although there are advantages in having *mundaukulu* located away from dwelling houses, farmers pointed out that there are negative impacts associated with these long distances, especially during harvest. The transportation of crops from these gardens takes more time than it would if the distances were shorter. Farmers are aware of the side-effects of positioning gardens in such a manner, but the problems faced are solved using local solutions. Group labour is sometimes used in a reciprocal manner to ease the burden of harvesting and transportation of the crop to storage sites, which are located in the villages. Farmers harvest and carry the crop harvested using labour sourced from several households. This labour is used collectively in successive gardens of the group members, until each member has his or her maize crop harvested and transported to the storage areas. Those who have an adequate income employ paid labour to assist in this process. The fact that food security is assured in this manner encourages farmers to accept the cost associated with distance, in terms of transport, as a requirement for achieving self-sufficiency in food. The farmers' decisions are rational and based on well-informed choices.

The strategy of the Malawi Government since independence in 1964, private seed companies, and some NGOs in recent times, has been to advocate the growing of hybrid maize as a measure that will ensure food security, based on scientific evidence which shows that the yield of hybrid maize is higher than that of local maize. However, hybrid maize has yet to be accepted by farmers as providing adequate food security, because it is susceptible to weevil damage soon after harvest, and is therefore difficult to store over the lengthy periods up to the next harvest which are required for maintaining household food security. Farmers assess and evaluate Western technologies, such as hybrid maize, using local varieties as a control to see if their 'traits' fit in their normal farming practices that ensure food security. Interestingly, very few farmers (3) planted more hybrid than local maize, while many more (108), either planted local maize only (55), or combined local and hybrid maize (53).

To store hybrid maize for these periods, it must be protected from weevils by using pesticides, which has a clear financial cost. This is considered unnecessary by farmers, because local maize can be stored over the same period without pesticide use. Moreover, farmers consider that the cost of pesticides is too high and therefore restrictive and prohibitive. Yet it is doubtful that, even if the price of pesticides is reduced, or even eliminated, farmers would reduce the area put to local maize and increase the area put to hybrid maize. Women in the focus group discussion agreed that, in any case, flour made from local maize lasted longer than flour made from hybrid maize of the same volume. Another farmer said that the taste of *sima* made from local maize flour is superior to that made from hybrid maize. This produced a lively debate on taste that extended to the taste of both roasted and boiled green maize. It was agreed that local maize is sweeter than most hybrid maize varieties. This shows that farmers' decisions are made after considering and evaluating a range of factors. For farmers, the yield of crops such as maize is not the only factor; crop choice also involves taste and the amount of the final product, such as *sima,* which can be made from it. It is now a known scientific fact that local maize makes tasty and more plates of *sima* from equal volumes of maize flour made from some hybrid varieties.[27]

[27]C. Chienga, Effects of Variety and Processing Methods on Physical and Chemical Characteristics of Maize Flour and Sensory Properties of *Nsima*, MSc thesis, University of Malawi, Bunda College of Agriculture, Lilongwe, 2012.

The importance of farmers having sufficient food within each household is shown by the fact that, although many crops (19) are grown in the study area, the dominant ones are food crops (Table 1.2). In addition, crops that are considered as important foods, such as maize, beans, mangoes, pineapples, bananas, cassava, sweet potatoes, green vegetables and sugarcane, are grown by all farmers. The patterns of crops grown indicate that farmers are able to satisfy nearly all their dietary needs from their own production. Maize and cassava are a source of carbohydrates, beans are a source of proteins, green-vegetables are a source of vitamins, and fruits are a source of minerals and important vitamins such as vitamin A. Other crops, such as cassava and sweet potatoes, have leaves that are cooked as green vegetables and their tubers used as a source of carbohydrates by eating them raw, roasted or cooked. The piths of maize plants that have empty cobs (no grains in them) are eaten as a snack for their sugar content while green after producing a tassel. The production of a tassel is a sign that the sugar in the pith is suitable for consumption because its taste is seen to be pleasant at this time. Eating the pith before the formation of a tassel is avoided by farmers as it is not considered to be sufficiently sweet. Moreover, some farmers reported that those who have eaten immature maize piths have developed a strange fever. This may be true, or it might be a way to scare children so that they do not eat maize piths before cobs fully develop to a stage that farmers can distinguish empty cobs from those full of grains. The practice of ensuring self-sufficiency in food from various gardens may reduce the intensity of production that leads to excessive nutrient mining common in conventional agriculture. It thus contributes to sustaining production over a long period without heavily relying on external inputs.

Table 1.2 Information on crops grown in the study area

Name of the crops (in order of importance and nature of use)	Smallholder farmers growing the crops	
	Number of farmers growing the crop	Percentage of farmers growing the crop
Maize	111	100%
Beans	111	100%
Pumpkins	111	100%
Mangoes	111	100%

Pineapples	111	100%
Bananas	111	100%
Cassava	111	100%
Sweet potatoes	111	100%
Green vegetables	111	100%
Sugar cane	111	100%
Tomatoes	100	91%
Guavas	90	81%
Avocado pears	60	54%
Peas	60	54%
Oranges	30	27%
Yams	20	19%
Tobacco	3	3%
Ground nuts	2	2%
Millet	2	2%
N=111		

1.3.2 Farmers' Soil Knowledge Systems

Farmers in the study area have a range of knowledges about the soils found across their three garden types. Farmers classify soils according to their fertility levels in order to deploy suitable and appropriate agricultural management practices. In assessing soil quality, all the farmers in the study classified them, firstly, in relation to the length of time of continuous cultivation of a given field; and, secondly, in relation to soil properties, primarily the soil's 'slippery' or coarse texture. The farmers' determination of soil texture resembles scientific ways of classification, which interestingly is also based on the feel method. Rubbing soil between the fingers to determine soil texture is both a scientific as well as an indigenous way of determining its properties. Soils in virgin gardens (mphangula), irrespective of their actual properties, are considered to be rich in soil nutrients before cultivation is undertaken. Farmers recognize that the natural vegetation of virgin land provides the soil with dry matter, which decomposes to generate high levels of soil fertility. Indeed, all respondents said that a good garden site must have *vundira,* which is dry plant matter that is decomposing, similar to peat or compost, and in short could be termed as humus. It was established during focus group discussions that farmers deliberately choose garden sites that have dense vegetation because this indicates high soil fertility in the form of *vundira*(humus).

Farmers recognize that the denser the vegetation, the higher the dry matter available that forms *vundira.* However, farmers disaggregated dense vegetation and ranked that of *Brachystegia* woodlands to be superior to any other vegetation as an indicator of higher soil fertility, and therefore good and effective pointer for crop production, especially maize. One farmer specifically pointed out, without prompting, that he is cultivating a garden 10 kilometres away from the village because that was where *munyozi (Brachystegia spiciformis Benth),* a *Brachystegia* tree species, was found. Unsurprisingly, farmers are increasingly concerned that there is a growing shortage of available virgin land (where they also obtain other resources such as mushrooms, edible insects and wild plants for local medicines) in the study area; all have noted that virgin land has become scarcer with increased population densities, partly as a result of people renting houses in the area who work in the nearby city of Mzuzu. These are also renting gardens from farmers in the area who then have opened up new farmland replacing fields rented out and reducing forest areas further. The expansion into newly opened farmlands has resulted in more area being put under agricultural activities. There is general agreement that: '*Sono mizi njina ndi malo ghamunda walero ghakusowa*', a statement made by an elderly man at a focus group meeting, which can be translated as 'there is very little virgin land left for conversion to farmland because of the increased number of villages in the area'. Indeed, even the agricultural extension worker in the study area has records that show that the population of farmers in the study area has more than doubled from its 1970 figure of 1000 households. In addition, there is evidence that each field assistant in the country is responsible for about 2900 farm families, implying there has been an increase in the farming families since 1970.[28]

[28]A. Langyintuo, Malawi Maize Sector Stakeholders' Workshop Report: Strengthening Seed Marketing Incentives in Southern Africa to Increase the Impact of Maize Breeding Research Project. International Maize and Wheat Improvement Centre (CIMMYT), Harare, Zimbabwe, 2004.

Farmers said that because of the increased demand for land for cultivation from those wanting to rent it, they are forced to choose virgin land for conversion to farmland, but not necessarily based on the preferred indicators of soil fertility, such as the presence of black soils and *Brachystegia species*. Economic gains from renting can influence well-established local practices in choosing cultivation sites. Farmers are forced to choose land that is less fertile as a result of these external forces, and the cultivation of such sites becomes unsustainable and less beneficial, as the condition of low soil fertility demands the purchase and use of chemical fertilizers to improve crop yields. This is a cause for concern for farmers whose income was found to have an annual mean of US $457 during the study in the 2006/2007 cropping season.

Scientists have established that *Brachystegia* is a nitrogen-fixing plant. Farmers are generally not aware of this particular scientific finding, however, and certainly may not understand the process of nitrogen fixing in the soil by the *Brachystegia,* but they are well aware of the high soil fertility levels in those areas where the species is dominant. The characteristic *Brachystegia* has to fix nitrogen in the soil therefore explains the high fertility levels that farmers attach to it through their observation of the growth of crops on those soils that have had such trees on them before conversion to farmland. The ways of knowing are certainly different, but there is a common fact, that is true to both scientists and farmers, which is that the site is rich in soil nutrients, especially nitrogen; one conclusion is achieved by observing crops grown, while the other is through the use of Western science and technology.

The farmers' knowledge associated with *Brachystegia* is likely to be gradually lost over time as a result of the natural habitat being slowly lost. It will be difficult for farmers to demonstrate to their children how to choose a good cultivation site for crops, particularly maize, in the absence of such a key soil fertility indicator. Indigenous knowledge is passed on between generations in several ways, but a key way is through observation. The old adage that 'seeing is believing' is borne out in this situation. Increasingly, future generations are less and less likely to see the species in its natural habitat; they will be less able to choose a suitable cultivation site based on the presence of *Brachystegia* and therefore in due course will not associate it with soil fertility. Although this may be the unavoidable outcome of the disappearance of *Brachystegia*, other soil fertility indicators, such as dark-coloured and black soils, may replace them as major soil fertility indicators.

Farmers are already using this indicator in *dimba,* which is a sign that indigenous knowledge changes and evolves with varying environmental and ecological conditions. In short, indigenous knowledge is adaptable to changing times and circumstances.

In-depth group discussions revealed that farmers see *vundira* as 'providing food for crops'. It is, therefore, not surprising that most farmers (94%) had more than one *mundaukulu,* which had been achieved through the process of converting as many sites as possible, which had *Brachystegia,* into farmlands. It can be argued that this is an adaptation of the former practice of shifting cultivation (that left tree stumps in the field for regeneration) that is now practiced by very few farmers in the study area. To maintain and improve soil fertility in cultivated fields, farmers' soil management practices revolve around the continued production of *vundira,* and, in gardens that are not *mphangula,* the incorporation of crop residues into the soil is used to maintain fertility. Hence, this maintains high crop yields for as long as this can be achieved after the benefits of *vundira* from virgin land have been exhausted.

The plant food provided by *vundira* is called *mchere* in the local language, which is also the name given to kitchen salt. Farmers understand that plant food is depleted over time by crops, resulting in decreasing yields, for each additional year of cultivation. Consequently, gardens 'lose the strength' to support crops over a period of time as a result of prolonged and continuous use. When the garden reaches such a state, the farmers describe this state as '*munda wasukuluka.*' This translates as the soil in the garden has lost its fertility.

Focus group discussions showed that knowledge about virgin lands was widely shared amongst farmers across all ages and both genders. All members of the groups actively participated in the discussions, and it was clear that knowledge that has a direct impact on the yield of food crops, and by extension food security, is well shared amongst farmers. This is mostly referred to as survival strategies by many scholars such as Chambers and Beckford.[29]

---

[29]Robert Chambers, *Rural Development: Putting the Last First*. Longman: London: 1983; C. Beckford, D. Barker and S. Bailey, 2007, "Adaptation, Innovation and Domestic Food Production in Jamaica: Some Examples of Survival Strategies of Small Scale Farmers," *Singapore Journal of Tropical Geography* 28, pp 273-286.

Farmers allow children from the age of five onwards to cultivate in *khonde* as a learning process through practice and observation. Subsequently, at thirteen years of age or more, which is considered to be an age when children are about to be adults, farmers allow them to have their own independent gardens away from *khonde*. The farmers emphasise that children at this age need to have their own gardens, not only for training purposes, but also so that they stop consuming green maize from the *mundaukulu*. It seems that children are associated with increased consumption of green maize, which is considered to threaten food security as the amount of maize to be consumed as *sima* is correspondingly reduced. Because *sima* is the staple food for households in the study area, any use of maize that threatens its availability is a threat to household food security. Furthermore, the children's gardens increase children's knowledge of farming, because they can observe their own farming practices and compare these with those in their parents' gardens. Children can then make informed choices. For farmers, the ability of the children to be independent and to be able to evaluate and assess farming practices on their own is a necessary process to go through, if they are to graduate as successful and responsible farmers. Indigenous knowledge becomes a tool for teaching and for the maintenance of desirable cultural norms, expectations and values within the society.

Farmers are also accompanied by children of all ages to the fields during almost all agricultural activities undertaken. Children make ridges, plant and weed maize in *mundaukulu* under the supervision of parents, for example. All this is done in order to give the children practical courses in farming. In the process, children acquire knowledge themselves through observation and practice. Children above thirteen are sometimes seen cultivating *mundaukulu* in the absence of parents, which is an indication that at this age they are considered to have acquired enough knowledge to carry out cultivation practices, such as weeding and banking, on their own. The responsibility of having part of a *khonde* garden, their own small gardens and participating in farming activities in the presence of parents and relatives, provides an opportunity for children to observe these operations and to learn through practice. Farmers are aware that without such practical lessons, children will not grow into successful farmers in the future. One farmer was especially frank and was of the opinion that children, who are not taught farming practices, become lazy and end up as thieves. This is a link that is very interesting, in terms of knowledge transfer in farming being associated with an impact on social behaviour, as this farmer linked excessive free time for children to the development of anti-social behaviour and laziness. Farming lessons are seen to become part of those processes that are used to make children acquire discipline and good manners. Furthermore, most farmers consider that those people who buy food for household consumption, especially maize in a year when everyone else has adequate surpluses from their own production, have received an inadequate farming training during childhood; such farmers are said to be lazy. Effort is, therefore, made to prepare children to be productive farmers in the future to avoid parents being scorned by fellow farmers as having failed to raise their children in a proper manner.

The act of the first weeding stage in maize cultivation is considered to be very important by farmers. Children between five and twelve years of age are frequently seen to be very close to where parents are weeding. The first weeding involves removing weeds from the ridge and placing them in the trough. Since weeds can be very close to the planting station itself, it is common for children to cut the crops in the process. To limit this damage, children are under close supervision. Parents instruct children to pull weeds out by hand that are either between maize plants in a planting station, or very close (2 cm or less), instead of using a hoe, which can cause serious damage to the young maize plants.

Farmers generally agreed that the length of time it takes a garden to become less productive, that is, to lose its 'strength to support crops', depends largely on soil texture. Farmers are, therefore, aware of the crucial difference between what they call 'slippery' soils and loose soils. They particularly noted that loose soils become less fertile more quickly with use than slippery soils, even when planted with similar crops. Many farmers (93%) plant bananas on ant hills which are considered to have 'slippery' soils. Banana clones planted on ant hills are said to produce bunches for up to fifty years or more, as shown by bananas planted in 1949 in the study area. Conversely, those bananas planted on the flat lands that have loose soils often die in the first year of planting unless manure is applied to the planting station. Furthermore, farmers also showed me sites that have loose soils around villages, which have now been planted with cassava within ten years since the first opening of the land for cultivation, because the production of maize has become unsatisfactory on these soils, with 30% of plants producing cobs without grain. In contrast, ant hills within such locations are still being put to maize and produce cobs full of grain (Figure 1.2).

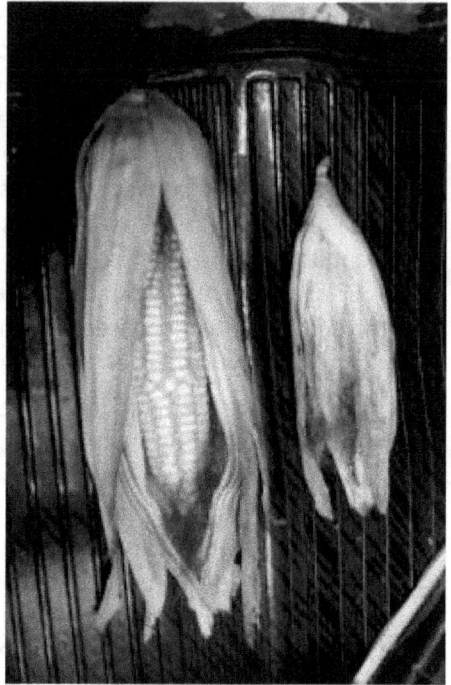

Fig. 1.3 Full and empty maize cobs

Table 1.3 shows that farmers of all ages prefer to plant bananas on ant hills, probably a factor that can be considered to show the effectiveness of the training process children undergo before they become adults. Children or young adults under 21 years of age plant on ant hills by imitating their parents.

Table 1.3 Information on smallholder farmers banana planting sites by farmers' age

Age category	Planting on flat	Percentage	Planting on ant hills	Percentage	Total	Percentage
<21	2	25	6	75%	8	7.2%
22-50	5	8.2	56	91.8%	61	55%
>51	1	2.4	41	97.6%	42	37.8%
Total	8	7.2	103	92.8%	111	100%

Farmers understand the loss of 'strength' of soil after use, by relating it to the similar effect that effort has on their ability to perform work. They get tired after working for a long period of time, and, in the same way soils also become 'tired' from continuous cultivation. The 'slippery' soils take a longer period of time to lose 'strength', in a similar manner that a strong person can endure a long period of hard work, while loose soils lose strength in a manner emulated by a weaker person who gets tired very quickly after working for a shorter period of time. The length of time taken for soils to lose 'strength' was emphasized by farmers by giving an example of 'slippery' soils, such as those found on ant hills. One farmer demonstrated this by showing an ant hill that has banana clones which were first planted in 1949. The banana plants were tall and were still growing vigorously; some had big bunches of fruits on them, a clear sign that the crops were still healthy and an indicator of maintained soil fertility. Many farmers emphasized this point further, by reporting that even the bananas that survived the first year on flat lands with 'loose soils' grow with less vigour and produce comparatively smaller bunches and fruits than those growing on ant hills. Farmers make detailed observations on a daily basis and they understand that bananas planted on the flat die because of lack of soil moisture soon after the rainy season is over and less nutrient availability in such soils. In order to retain moisture and introduce additional nutrients on these soils, farmers dig pits one metre deep and one metre wide, which are then filled with manure or plant matter that later decomposes to provide nutrients to bananas planted in these pits. The pits have an additional role of collecting rainwater, thereby retaining the moisture over longer periods of time than is normal for flat lands in the area. This generally sustains the growing bananas up to the next rainy season. Farmers understand that successful water retention is based on the pits' capacity to hold water, which results from the manure applied, which, in turn, retains soil moisture more effectively as a result of improved soil structure and texture.

Those farmers who cultivated a relatively smaller total land area, around 1-2 hectares, planted more maize on ant hills than bananas and pumpkins. They trade off space by making full use of the fertile soils on ant hills to increase their staple food crop which is maize. Some are forced, nonetheless, to plant bananas on ant hills, despite their limited farm size, because where termites (*Reticulitermes species)* are active on ant hills, they can destroy maize plants by attacking the stem just above the ground, thus killing the plant. However, termites are unable to destroy bananas because they have large, multilayered stems, and termites tend to feed on the outer layers that are dead and dry, so leaving the plant without damage.

Soil samples were collected from both ant hills and flat lands, and were analyzed to compare the farmers' ways of understanding soil texture and fertility with those of Western science and technology. The samples were analyzed for soil texture, pH, nitrogen content and available phosphorus. The results of the findings are presented in Table 1.4.

Table 1.4 Information on soil properties found on ant hills and flat lands

Soil properties	Minimum		Maximum		Mode		Mean		Median	
	Ant hill	Flat	Ant hill	Flat	Ant hill	Flat	Ant hill	Flat	Ant hill	Flat
pH	5.500	4.300	7.400	6.100	6.600	4.900	6.373	5.180	6.400	5.300
Nitrogen %	0.115	0.070	0.285	0.185	0.115	0.070	0.218	0.120	0.218	0.110
Phosphorus (ppm)	2.123	2.622	10.740	11.928	3.144	2.622	4.169	5.509	3.523	3.901
Clay %	40	30	83	47	67	40	68	40	70	40
Sand %	7	23	27	50	10	40	15	39	13	40
Silt %	10	7	23	27	20	20	17	20	17	20
N=30										

Soils from ant hills have a higher clay content, with a mean of 68% compared to 40% on the flat lands. The mean sand content was 15% in the soils from ant hills, which was less than half of the mean sand content

(39%) in soils from the flat lands. The mean soil pH on flat lands was 5.2 and that of ant hill soils was 6.4. The mean nitrogen content of soils on ant hills was higher (0.22%) than the flat land soils (0.12%), but the mean available phosphorus in soils was higher on the flat lands (5.5 ppm) than on ant hill soils (4.2 ppm). The results confirm that the soils on ant hills have a higher soil fertility compared to those on the flat, confirming that 'slippery' soils are richer in soil nutrients than loose soils.

The differences in soil characteristics can be explained using Western science understandings. Soils that have a high clay content tend to bind phosphorus within them because of the adhesive properties of the clay. This explains the lower available phosphorus found on soils from ant hills compared to those found on flat lands. The high nitrogen content can be explained by the fact that ant hills are built by termites, and are produced by gathering plant residues on which the termites feed, and from which Nitrogen is released through the decomposition process, normally done by microbes such as bacteria in the soil. The process of gathering and decomposing plant residues by termites in ant hills, plus the metabolic wastes from termite digestive systems, therefore concentrates nitrogen in soils found on ant hills making it higher than those soils found on flat land.

The understanding of the significance of soil texture and fertility by farmers is comparable to that of Western science. Both groups understand the roles that soil particles play as soil texture determinants. They also understand that soil fertility is a function of the nutrients contained in soils; farmers are aware of the fact that soils without such nutrients are less fertile. Whereas Western science and technology analyses and establishes the exact content and nature of such material in soils, farmers establish the existence of such materials that make soils fertile by observing the vigour of crops grown on soil sites; for example, soils with a high nitrogen content make plants grow with vigour and produce a dark green colour in the leaves.

The significance of soil feel, which Western science sees as soil texture, is understood by farmers, primarily as an indicator of water holding capacity. Slippery soils are considered to hold moisture for longer periods of time after the end of the rainy season than the 'loose' coarser soils, which quickly become dry. This accords with the scientific fact that sandy soils lose water faster than clay soils. For example, pumpkins, and a few broad-leaved weeds, are frequently still green on ant hills in August and beyond, long after the end of rainy season (Figure 6.3). The characteristics and properties that farmers attach to 'slippery' soils, that they hold moisture for a long time after the last rains, is substantiated by the high clay content in ant hill soils.

Figure 1.4 Pumpkins on an ant hill are still green in the dry season

Nearly all farmers (96%) planted pumpkins on both ant hills and flat lands, explaining that pumpkins were planted on ant hills to extend their growing time beyond the rainy season. It becomes clear that the importance of the moisture holding capacity of clayey, 'slippery' soils cannot easily be ignored in the farmers' cropping practices. For farmers, soil fertility and the water holding capacity of soils go hand in hand, to the extent that they are essentially of equal importance. Farmers' agricultural management practices therefore are designed to benefit from the essential soil proper-ties of fertility and water retention capacity.

Pumpkins planted on 'loose' coarser soils die soon after the rainy season ends. The growing season of pumpkins on such soils, therefore, ends at the onset of the dry season at the end of April. Farmers' needs for green vegetables go beyond this time of the year; consequently, they utilize the fertility and water holding capacity of ant hills. The green vegetable and fresh crop production is further extended by the cultivation of *dimba* where residual moisture is available for crop production after the rainy season.

The soils in *dimba* and *njota* are much darker in colour, even black in some cases. These soils have characteristics similar to those soils found on the ant hills. They easily form a round ball when wet, are slippery when rubbed between fingers, and, most important of all, are very effective at retaining soil moisture, so necessary for plant growth. Farmers associate the dark colour of these soils with higher levels of soil fertility. Hence, *dimba* and *njota* are considered fertile because of their soil colour, and also because of the thick natural vegetation found on them, mainly composed of reeds and grass. It was established during focus group discussions that reeds are a key indicator of high soil fertility levels where they are found in abundance, and especially so if their stems are bigger in circumference than average human fingers.

Significantly, farmers identify clear differences between *dimba* and *njota*. The soil characteristics are similar, and, in some cases, they may even be identical. The real difference is the level of the water table during the rainy season. Farmers explained that both *dimba* and *njota* experience flooding in the rainy season. However, *njota* behave almost identically to upland soils during the rainy season. The soils in *njota* retain adequate moisture for plant growth and quickly drain excessive moisture. This makes them suitable for crop production such as maize, as the crop does not become yellow, which would be the case if water was retained so reducing soil aeration processes. The amount of the retained moisture allows soil aeration and is, therefore, sufficient for crop production in the rainy season. However, soils in *dimba* retain excessive moisture in the rainy season, resulting in water-logging which is unsuitable for some crops, especially maize. Water-logging in most cases results in some crops turning yellow, wilting and dying. Drainage during the rainy season to achieve a moisture content suitable for crops becomes difficult to impossible because of the high water table. In fact, farmers noted that some *dimba* become inundated in the rainy season, sometimes for periods of up to three months. Thus, they become wetlands under these conditions, with shallow waters that make crop cultivation at such times impossible.

1.3.3 Indigenous Soil Improvement Practices and Fertilizer Use

Farmers deploy agricultural management practices that can improve soil fertility, or at least retain it for as long as they cultivate a site. Soil improvement practices and fertilizer use begin from the first year of cultivation of a given field. The processes of cultivating a virgin upland involve land clearing that includes cutting trees using hand held axes and slashing grass using home made *pangas* (cutting knives). When trees are cut down or felled, big branches that can be used as firewood are gathered and transported home or sold to people who mould bricks in the study area or elsewhere. The smaller branches are laid down along with the slashed grass. These are allowed to dry over a period of about two to three months or more, and then set on fire. The burning is done on a calm hot day to create optimum conditions necessary for effective burning and sterilization of the soil. The farmers understand that by the end of two to three months of hot sunshine in the dry season, the branches and grass are dry enough to be fully burnt to ash. The ash created is then incorporated fully in the soil by tilling the land as a source of soil nutrients, especially potash.

All farmers agreed that the most suitable crop in the first year of cultivation of most *mundaukulu* is finger millet as the dominant crop in mixed stands. Indeed, two farmers who had converted virgin land to farmland in the year of the study planted it to finger millet. However, some farmers cultivate maize as the dominant crop in such mixed stands on virgin lands, especially when they have a pressing need for the major food crop. The main reason for preferring finger millet as a major crop in the mixture in the first year is that weeds are destroyed by burning. This is a very important factor for farmers since the weeding of finger millet is done by pulling weeds out of the soil by hand. Weeding for other crops such as maize is typically done using a hand-held hoe. The more weed seeds that are destroyed by burning, the fewer the weeds, and hence the less labour that has to be deployed, and the task of weeding becomes easier to undertake. In addition, ash is considered by farmers to be an effective plant food for finger millet, and so it tends to be grown only on newly converted farmlands that have been recently burnt. Ash contains potash, a necessary element for plant growth, and this explains the improved performance of millet under such conditions. A further reason put forward by farmers is that burning delays the subsequent appearance of weeds, such that the finger millet emerges on clean ground, at least when burning has been properly done. The crop, therefore, has limited competition from weeds for plant food, and especially so in its early growth stages when it is most vulnerable to such competition. This is then easily maintained through the growing season by weeding, when needed, especially because millet cannot be easily differentiated from other grasses in its early stages from emergence to about one month in age. Farmers noted that even when millet can be recognized, pulling out weeds that are at the same growth stage results in the potential destruction of the millet crop. Farmers avoid this crop destruction by 'sterilizing' the planting sites using fire.

The preparation of branches and slashed grass for effective burning is considered a practice that requires special knowledge, skills and talent in laying the branches and grass so that all branches completely burn into ash by the end of the process. Farmers lay branches in positions determined by their size. The smaller branches, with diameters of up to 3 cm, are put around bigger branches with a diameter above 3 cm, so that the smaller branches generate adequate heat to burn the bigger branches; it is the subsequent burning of the bigger branches that generates sufficient heat that is effective in killing weeds. Some pointed out that this knowledge is accumulated with experience, associated with age as a result of doing this process many times. The experienced farmers said that to attain effective high temperatures, the environmental conditions at the time of setting the fire and the whole period of burning need to be calm, with little or virtually no breeze present. Experienced farmers have the ability to forecast and identify such days using local knowledge and understandings of weather and this forms part of the skill-set they have; calm hot days are characterized by 'malawi', which is hot air seen in the form of waves rising from the surface of the earth, especially observed in cleared spaces such as gardens, roads or footpaths. Significantly, the temperature in the dry season, especially in the months of September and October, is above 27ºC, creating ideal conditions for effective burning. Although it rains throughout the year in the study area, the months of August, September and October experience relatively lower rainfall compared to the other months of the year. These climatic factors of low rainfall and high temperatures create ideal conditions for successful burning. The knowledge about when to set fire to the branches is based on the experience these experts have had with high temperatures in these months. Those who do not use such experts to burn branches fail to attain the killing effect of the resultant fire as insufficient heat is generated during burning, so that many weed seeds survive. The soil remains inadequately sterilized so that weeds appear in abundance as the millet emerges. This increased weed population creates a need for an increased labour demand during the weeding of the crop. An increase in weed population can result in the destruction of the emerging millet crop particularly when weeds are being uprooted. To avoid inadequate sterilization, many farmers said that they seek assistance from experienced farmers in the form of reciprocal labour.

Where burning has been effective, weeds appear relatively late in the crop growth cycle and are easier, therefore, to differentiate from the finger millet, especially if they are of the grass family. Furthermore, farmers are of the view that burning loosens the soil, which then requires less effort in the process of uprooting weeds. This point was raised as a very important advantage of effective burning. Farmers try as much as possible to make their work as simple and easy as possible, and, in so doing, they save energy for other activities. This may not be surprising as they have many crops located across several gardens to attend to. This behaviour shows that farmers are rational in allocating scarce resources such as labour. They plan in advance in order to rationalize such resources.

Although the farmers in the study are aware of the cultivation practices associated with finger millet, only two farmers grew the crop in the year of the study. However, most of them had grown finger millet when their gardens had been virgin lands, and therefore they retained a memory of the advantages of burning and the resultant ash in such fields. In addition, farmers tend to share labour in groups during cultivation activities, such as weeding in particular. Therefore, although few farmers cultivate finger millet at the moment, because of the limited availability of virgin land, weeding the crop on established land involves many families in the form of reciprocal labour sharing. The reciprocal labour sharing and previous cultivation of the crop by many farmers may explain the wide and extensive knowledge that farmers showed during focus group discussions on finger millet cultivation and the soil fertility management practices associated with it, despite the fact that most of them were not cultivating the crop in their own gardens in the year in which the study was conducted. Knowledge therefore is learned, renewed, shared and maintained through activities and practice. Organized group labour ensures knowledge is not easily lost or forgotten.

After the first year of cultivation on virgin land, maize then becomes the first choice major crop, and maize dominates the crop mixed stands for as long as the soil can support yields that are reasonable for the farmer. This varies amongst the farmers, but when 30% of the cobs have no grain, most farmers consider this to be unacceptable. Their understanding of yield at this stage is in terms of empty cobs per given number of plants along a ridge. During harvest time farmers first use a number of *mukukwe* (see Figure 1.1) made in their *mundaukulu* to measure yield and later use baskets as they carry the crop for storage in granaries. Quantities are measured in volumes that farmers understand and relate to measuring instruments such as baskets and plates.

Farmers do not have weighing scales to measure yield in kilograms, such that the available alternative to measure food for consumption is the baskets used to carry maize after harvesting from gardens to granaries. This is a convenient way for farmers to measure food for consumption, because farmers have baskets that are used for other uses, such as grain storage, before taking grains to a maize mill. More important to farmers is that the baskets are also used to take maize for consumption from granaries, and, as such, are used as a measure of the amount of food baskets consumed per month for a given family. Consequently, farmers measure yields of maize in terms of bags produced, numbers of full baskets carried from a given field and the level of maize cobs in the grain stores. This is done by counting layers made of cobs across the surface area of the granary. This gives farmers a clear indication of the volume of maize they use each time that maize is collected for consumption from the granary, by counting the remaining layers. Each layer is then linked to a time-scale in terms of the months it can take to exhaust the contents of the granary. Farmers note that it is then easier to estimate when grain is likely to be used up.

Once 30% of cobs are without grain (see Figure 1.2), the garden is either immediately put to fallow or is planted with cassava for up to two years or more, before being put to fallow. However, farmers who are employed and have additional income are able to improve yields through the application of chemical fertilizers, and therefore prolong crop production beyond what would be possible without chemical fertilizers, which is typically understood by farmers to be 10 years on sandy soils, and up to twenty years or more on dark soils.

39

Farmers' understandings of soil fertility loss is conceptualized along the lines of food preparation practices, and particularly the application of kitchen salt to side dishes made of vegetables, meat and beans, which are usually served with *sima*. These side dishes, in most cases, are prepared to last for more than one meal because they take time to prepare, and a considerable amount of firewood is used in the process. Consequently, their preparation beyond a single meal is considered to be a rational decision in order to save labour and firewood. Salt is applied to these dishes to the required taste. However, the desire to take the side dishes while warm during all meals requires that water be added in order to heat up the side dishes. This additional water dilutes the salt content, and so it becomes necessary to add salt to the dish to maintain its previous taste.

Farmers use this understanding behind salt behaviour in the consumption of side dishes to interpret soil fertility loss in gardens. The starting point in understanding their reasoning can be seen in the name farmers give to chemical fertilizers. Fertilizer is called *mchere*; and *mchere* is also the name given to kitchen salt. The importance of kitchen salt is seen in its properties of improving the taste of food to which it is applied. The salt content of such food becomes diluted as a result of water addition when it is heated up, and this understanding is applied to represent the similar effect of rain in adding water to soils in cultivated gardens which dilutes soil nutrients. However, this is seen to be different from the leaching of soil nutrients by rainwater, as this results in soil nutrients being washed deeper into the ground. Farmers believe that not all soil fertility loss can be explained by crop nutrient uptake, soil erosion and leaching. For farmers, it is clear that leaching cannot occur in pots that do not leak, and yet salt which is added to side dishes still fades in taste over time with each application of additional water. This fact then provides farmers with an analogous explanation as to how soil nutrients are lost in their gardens. This has a bearing on how farmers understand soil fertility in a somewhat different manner that goes beyond soil properties and texture. It also has an important bearing on land management and improvement practices that they undertake to maintain soil fertility in their gardens. Significantly, it is also different from the main scientific interpretation of soil fertility loss, which is based on factors such as leaching, crop uptake and soil erosion, for example, of which the farmers are also aware.

The soil, therefore, is thought to show the properties of both a non-living organism and of living things, as it becomes 'tired' and 'loses strength'

40

with continuous cultivation. The conceptualization of soil, both as a living and a non-living entity, might be incomprehensible to Western scientific ways of knowing. Yet this is the smallholder farmers' way of knowing and they move between these scenarios with ease. Both the living and the non-living nature of being are used to explain events that cannot adequately be explained by only one of these two when used independently. The farmers, therefore, try to be as holistic as possible in explaining their understanding of the nature and environments, in this case soil fertility, on their farms.

Farmers put gardens in uplands, particularly *mundaukulu* that have loose soils, to fallow for a minimum of one year after continuously cultivating them for a period of up to twenty years or more, depending on how many *munda* a given farmer has. Those with many *munda* 'rest' them for longer periods. For example, one farmer with four gardens has a *munda* that has been fallow for more than ten years (Figure 1.0).

Putting land to fallow is the least desirable option that farmers have for soil fertility maintenance, because it has a negative impact on food security, as it decreases the land area available for food production. Consequently, those with fewer gardens, and lower hectarages, are unable to put land to fallow for the longer periods of time necessary to replenish soil fertility. There is an additional reason for this reluctance by some; land left fallow is sometimes encroached on and used by relatives, and this can result in unnecessary and unpleasant family disputes. To avoid such disputes, cassava is often planted on land considered suitable for putting to fallow, as a means of retaining undisputable ownership of the land, but without being a significant drain on fertility. Indeed, cassava is thought to restore soil fertility in a similar manner to fallow. The crop is deliberately left to grow undisturbed with weeds, as minimum to zero weeding is done in the second year after planting. Weeds mature, die and decompose in the garden, replenishing soil fertility. This is possible because cassava can be left with weeds without yield being reduced. It is a crop that is harvested only when it is needed for consumption; otherwise it is stored in the garden as a living plant, while utilizing only its leaves as a green vegetable. It can, therefore, be left in the field for periods of time that can extend over two years or more, replenishing soil nutrients.

As fallow is undesirable or even impossible for some farmers for soil fertility maintenance, they opt to use chemical fertilizers to improve crop yields. They explained that, although fertilizers are effective in raising crop

41

yields, particularly of maize in their *mundaukulu,* which have low levels of soil nutrients, the cost is high and therefore restrictive, and especially so considering that almost all food produced is for household consumption. Thus, many farmers are obliged to engage in employment, partly to generate the funds to acquire chemical fertilizers. Some 48% of farmers are engaged in some form of employment, ranging from casual work, known as *ganyu,* to full employment. *Ganyu* refers to the selling of labour by doing paid piece-work. Most farmers said that money raised from employment is primarily for the purchase of chemical fertilizers.

Indeed, all farmers in the study area use at least two bags of fertilizers on some of their *mundaukulu,* because most farmers cannot afford to apply fertilizer to all their cultivated areas due to its high cost (65 $ per 50 kg bag) and their corresponding very low incomes (mean of $457 per annum). Most fertilizer is accessed mainly through a government-subsidized input scheme using coupons. Each farmer is entitled to fertilizer coupons that enable him/her to buy one 50 kg bag of basal fertilizer and one 50 kg bag of top dressing fertilizer. Those who have higher incomes and are employed use the recommended rate of fertilization, which is at least four bags of basal dressing and two bags of top dressing per hectare. The use of fertilizer by farmers is also supported through social bonds and obligations. This study found that those who are unable to purchase chemical fertilizers, because of inadequate income, often engaged either in 'unpaid' labour in exchange for fertilizer, or simply begged from friends and relatives. There are social obligations within the society, where the needy are helped by those who are considered to be blessed with 'abundant' resources. Fertilizer, just like labour, was sourced 'free' of charge from friends and relatives, not necessarily as a future stock of capital to be drawn upon later, but a mere social obligation based on culture. The idea of the 'genuine' needy deserving assistance is a norm, value and expectation that is taught right from childhood, and which is followed to the letter. The village heads and their councillors, called *nduna,* ensure this is attained by 'reprimanding' those who fail to help but without any form of punishment. Giving to the needy is not enforced through punishment, but through persuasion and encouragement in an orderly manner that ensures little resentment from those with resources to give to others.

However, farmers are convinced beyond doubt that chemical fertilizers 'burn' the soil. Farmers apply chemical fertilizers to crops only reluctantly because they are aware of their side-effects. Focus group discussions showed that farmers' past experiences with chemical fertilizer use have been characterized by mixed fortunes, and this has made them associate chemical fertilizers with the disturbance and destruction of the natural mechanisms of soil nutrient replenishment. Some farmers made it very clear that farms which had once been applied with chemical fertilizers have produced less than half of the yields realized without fertilizer use, when chemical fertilizers have not been applied in subsequent years. Meanwhile, those farms who have not been exposed to chemical fertilizers have maintained yields over subsequent years, with output loss of less than a layer in granaries each year, probably an indication that chemical fertilizers indeed have a negative impact on soil microbes that help in the fertility maintenance.

The observations which farmers have made show that once chemical fertilizer is applied to a field, yields of a given crop can only be maintained with the continued use of fertilizer in subsequent years of cultivation. However, gardens that have had no previous application of fertilizer are known to produce comparably higher yields of crops, especially of maize, than those gardens that have had only one chemical fertilizer application and none thereafter. This then is considered a burden for farmers, and they try as far as possible to maintain at least one garden where chemical fertilizer is not applied to crops. Many farmers recalled that in the past, particularly between the 1960s and 1970s, crops were grown without the application of chemical fertilizers and yields were adequate for meeting family food needs. Farmers compare the results of their crop performances over a long period of time, that, in some cases, can be beyond forty years. This is a demonstration of a high level of memory, which may explain farmers' lesser use of written records about their farming practices.

Interestingly, it has been scientifically established that some chemical fertilizers, such as sulphate of ammonia, make soils acidic. The change in soil pH can affect and at times can upset the microbial activities in the soil that decompose dry matter.[30] In Malawi, sulphate of ammonia fertilizer is now only recommended for application to irrigated rice, where its acidifying properties are beneficial to the crop; it has been withdrawn from being applied to upland maize and other crops grown in the mixed stands. It

---

[30]L.B. Taiwo and B.A. Oso, "Influence of Composting Technologies on Microbial Succession, Temperature and pH in a Composting Municipal Solid Waste." *African Journal of Biotechnology* 3(4), 2004, pp 239-243.

appears that the farmers' understandings of the side-effects of chemical fertilizers on soil are similar and in agreement with those of scientists. Some farmers argued that they observed this phenomenon long before scientists had linked chemical fertilizers to such side-effects.

Farmers have discovered another problem associated with chemical fertilizer, in particular, urea, which is yet to be verified using Western science and technology. They all agreed that urea makes beans' leaves wilt and drop off the plant, and eventually the plant dies. The common crop mixed stands of beans and maize, therefore, is now managed differently by those farmers who use urea. It is a normal practice to plant beans and maize in the same planting station (hole) along ridges about 90 cm apart (the ridges are also about 90 cm apart). However, because of the problem of urea use for beans, to avoid the loss of bean yield, farmers now plant beans in different planting holes (stations) from maize. Urea fertilizer is then placed using a fertilizer cup in a hole made in the ground with a sharp stick 6-10 cm from the maize plant, and up to 30-60 cm away from the beans. At this distance, the side-effects of urea are not 'seen' on the bean plants. The distance is therefore considered safe and adequate, based on the lack of visual side-effects on the bean plants. Obvious visible effects on plants, such as wilting and eventual death of plants, are very important to farmers, as decisions are based primarily on what they are able to see.

As we have seen, farmers cultivate ant hills for long periods of time, close to fifty years in some instances, without applying chemical fertilizers to them; bananas planted on ant hills are cultivated typically without the application of chemical fertilizers (Figure 1.4) The soil fertility is maintained by leaving stems and leaves, from which banana bunches have been harvested, to decompose within and around the clones. Such practices are considered adequate to maintain soil fertility, as banana yields do not decrease for as long as these practices are maintained over their growing period.

Figure 1.5 Bananas on ant hill

In *mundaukulu* and *khonde*, crop residues and weeds are buried under ridges constructed for planting crops in the subsequent year of cultivation. This practice is replicated each year of cultivation when ridging is done up to the time that the garden is put into fallow. The second weeding then involves rebuilding the ridges on which crops are planted, by taking soil from the troughs and placing it on the existing ridge using a hand held hoe. This buries weeds in the ridge that is being rebuilt. This practice is technically known as banking. Soil fertility, therefore, is maintained by weeds and crop residues which eventually decompose within the ridge, releasing plant nutrients in the soil. This annual incorporation of weeds and crop residues to maintain soil fertility resembles, for farmers, the need to add salt to side dishes to maintain their taste.

Some of the crop residues and weeds are gathered on particular sites, that are deliberately and carefully chosen, and then burnt. These sites are then used as pumpkin planting stations by making a hole in the ash where pumpkin seeds are placed. Ash is utilized as a fertilizer for pumpkins and is an important element in soil fertility improvement as it contains potash, an element which Western science and technology has associated with quality in many crops, such as tobacco, for example. Most farmers (94.6%) planted pumpkins on ash (Table 1.4) and the practice was undertaken across all age groups. The planting sites are chosen based on positions that limit the effects of pumpkins on reducing the growth of other crops that are smaller, such as beans, by reducing access to sunlight. Pumpkins have bigger leaves, especially compared to those of beans and maize, and they tend to grow very quickly from the third week after emergence compared to other crops. Farmers said that the choice of pumpkin planting sites involves consideration of these physical characteristics that include the large pumpkin leaves and their shading effect on other plants.

Table 1.4 Information on smallholder farmers' pumpkin planting sites by age

Age category in years	Planting without ash	Percentage	Planting on ash	Percentage	Total
<21	1	12.5%	7	87.5%	8
22-50	3	4.9%	58	95.1%	61
>51	2	4.8%	40	95.2%	42
Total	6	5.4%	105	94.6%	111

Planting pumpkins on ash is a deliberate strategy because ash improves the taste of pumpkins, a typical quality aspect associated with potassium, as well as improving the vigour of the plants during growth. It is not surprising, therefore, that farmers tend to plant pumpkins on ash. However, taste is not only related to using ash as a fertilizer; most farmers also consider that taste is a function of variety and type of pumpkin. However, within a given type and variety, such as Autumn King, which is locally known as *jungu*, it is believed that taste is improved when these are grown on sites that have ash.

Pumpkins are sometimes called runners, because they spread stems along the ground. Since the combination of big leaves and the growth by spreading along the ground are seen to inhibit the growth of other crops in the field, pumpkins are, therefore, planted around the periphery of the gardens, with just one or two plants per ant hill. More than ten pumpkins per hectare is considered very high for the successful growth of other crops in the field. The side-effect of pumpkins on other plants is not only understood in terms of competition for the light necessary for photosynthesis, but also their physical nature of almost growing over smaller crops. Farmers are careful and limit this effect of pumpkins on other crops by uprooting branches that grow over other plants and then laying them in the furrows.

The practice of using crop residues to maintain soil fertility is very widely used in *khonde* gardens. Crops, such as beans, after being harvested in *dimba* and *mundaukulu,* are processed at home. Their dry matter waste, including stems and bean pods from processing, are laid down along ridge troughs in *khonde* gardens to decompose. The processing of maize to make white flour for *sima* is done by pounding the grain using a mortar and a pestle after the husks and the cob have been removed. Pounding breaks the grain into the three parts, which are the endosperm, bran and the embryo (or germ). The endosperm is ground into white flour, while the bran and embryo become waste products of this process that are also put in ridge troughs around *khonde,* if they are not fed to livestock such as chickens or cattle. Even when they are fed to livestock, the remnants are later swept onto the ridge troughs in the *khonde* gardens.

It is important to note that maize flour can also be made from whole grain, and, in such circumstances, very little waste is left for feeding livestock and maintaining soil fertility. However, most farmers said that they prefer white flour, mainly because of the improved taste of *sima* prepared from it, in comparison to the *sima* made from whole grain flour. There is of course the added advantage that the waste products can be used to improve soil fertility or fed to livestock.

Farmers do not see the application of crop residues to *khonde* as anything special or remarkable. The *khonde* are near to the house where the processing of crops is done, and hence the waste is dumped as close as possible to the site. Hence, such activities have become part of everyday life, and farmers perform them without much attention paid to the reasons underlying their actions. However, during in-depth discussions and further probing, it was said that some crops, particularly maize, when grown with such dry matter decomposing in the soil, grow with vigour and produce bigger cobs, compared to similar crops grown without decomposing dry matter being available. Farmers have a clear understanding of the effects of decomposing dry matter on the growth of crops from long-term observations of crop performance in *khonde* gardens in particular.

Farmers who have livestock, such as goats, allow them to feed in *khonde* and *mundaukulu* in the dry season. The droppings from such animals add nutrients to soils, although they remove some through feeding. It is a common practice to apply manure from goats, chickens and cattle to *khonde* and *mundaukulu* to improve their fertility. Manure can be bought from those farmers who have livestock, or sometimes is given away free when kraals are cleaned out. Manure can pose a threat to animals' health, particularly in the rainy season. Disease vectors, such as houseflies, breed in such places and farmers are keen to reduce the breeding grounds of house flies by allowing fellow farmers to collect manure from kraals to apply in their gardens.

Soil improvement practices in *dimba* are complicated. Farmers largely avoid using chemical fertilizer in these areas, as was made clear in focus group discussions. The few that use chemical fertilizers in *dimba* tend to apply them to high-value crops, such as tomatoes and other vegetables that are sold for cash. Even under such conditions, farmers seem to be more eager to use chicken droppings than chemical fertilizers. *Dimba* soils are darker in colour, as compared to many upland soils that are red *katondo* soils (Figure 1.5). Dark coloured soils are considered more fertile than the red *katondo* soils, and hence the crops in *dimba* grow with more vigour compared to those on upland soils. The farmers note that maize in *dimba* rarely produce cobs without grain, even after cultivating in these sites for a long time, such as over twenty years. Farmers are also aware that top soils from upland gardens are deposited in *dimba,* and this replenishes soil nutrients so that the application of chemical fertilizers to these soils is a luxury farmers with low average incomes are unwilling to undertake. Farmers show that economic factors are considered in deciding to use chemical fertilizers by applying them only to crops that they sell.

To maintain soil fertility, farmers leave weeds to grow in *dimba* during the rainy season, even when they have perennial crops, such as bananas and sugarcane already growing there. The farmers simply slash grass that is around such crops, let it dry and then burn it away from the perennial crops, thereby adding potash to soils.

Farmers said that weeds left to grow during the rainy season have an additional purpose, that of slowing down rain water so that the sediment from upstream is deposited in the *dimba*. This helps to replenish fertility that is taken up by crops cultivated in *dimba*. Farmers are aware that farming activities in *dimba*, such as land preparation and tilling, loosen the soil and expose it to erosion by rain water. Tilling weakens stream banks,

Figure 1.6 Dark coloured dimba soil

and drains made to lower the water-table can be unstable. Soil erosion widens stream and drain width as walls collapse, and this can have a direct impact of reducing the area for crop cultivation. This is certainly considered unacceptable by farmers, especially because some *dimba* tend to be narrow enough anyway, at between 10-20 metres in width, divided in the centre by a permanent stream. Any increase in the width of the stream results in a reduced area for crops.

Farmers clear their *dimba* in the dry season from April onwards by slashing and burning, as described above. All plant matter that resists burning is buried under the soil to decompose for the release of soil nutrients. The farmers make no effort to create high temperatures from the fire for effective burning in *dimba,* as they do in the finger millet gardens, for example. The plant residues that do not burn completely are simply ploughed into the soil through tilling, as is the ash. This is not surprising, as a high, intensive heat is unlikely to be attained largely because the soils in *dimba* are wet at this time of the year from residual moisture, and the complete drying of dry matter is therefore difficult to achieve. Furthermore, waiting for the complete drying of these residues would result in the delayed planting of crops, such as maize, which takes 90-190 days to mature and to be dry enough to be harvested. This can result in such crops either being destroyed by the rains in the rainy season in December, or being heavily affected by mould infestation. Therefore, farmers are careful to start cultivation in sufficient time to allow maize crops to mature and to be harvested before the onset of the next rainy season. This may also be the explanation behind farmers' reluctance to plant finger millet in *dimba,* despite the ash being available as a fertilizer in each year of cultivation. Finger millet requires a longer growing time than maize to mature, to completely dry out and to be ready for harvesting.

Cultivation practices in *dimba* are different from *mundaukulu* and *khonde*. Crops are planted on the flat in *dimba,* by making a hole using a hoe and without making ridges. Planting is similar to the cultivation practice of finger millet production, which is planted by broadcasting the seed on the flat that is then immediately lightly tilled in using a light hoe. Planting without making ridges is also practiced on ant hills. It appears that farmers are aware of the moisture regime in the *dimba* being of particular importance for plant growth. For farmers, soil improvement in *dimba* is more a question of managing the moisture regime than the soil fertility itself. Farmers are aware that planting on ridges would increase the distance between soil moisture and the crop roots, as ridges have a function by their very nature and shape to drain excessive moisture that is unnecessary for plant growth. In *dimba,* this function of the ridge is unnecessary and can be problematic, as moisture available to plants is residual. Farmers place seeds in planting stations as close as possible to the source of the residual moisture so that the plant roots are able to access the soil moisture as the water table continues to lower from the effect of the dry season progression. Planting on the flat makes it easier for plant roots to grow up to the depth at which the residual moisture is available in the soil.

Furthermore, farmers have observed that when ridges are made in *dimba,* the soil is drained to the extent that the soil on top of the ridge becomes so dry that its dryness is readily observable by eye, even while the trough is wet. This makes it necessary to apply water to crops to supplement the residual moisture. Farmers' understandings of drainage are, therefore, related to the distance between the roots of crops and available soil moisture, which reflects their accumulated knowledge of water-table levels in the *dimba*. They are aware that crop roots have a finite depth of growth, and, beyond that depth, they are unable to reach the moisture in the soil necessary for growth. Consequently, plants wilt and require supplementary water supply. Farmers recognize that the water-table is important for crop cultivation in *dimba,* and it is therefore carefully managed in a manner that reflects its importance by cultivating on the flat. When the water level goes beyond the plant root zone, as a result of factors beyond the farmers' control, such as dry spells during some rainy seasons, farmers may apply water to plants.

The application of additional water to crops, however, in order to supplement residual moisture, is viewed as creating an unnecessary labour demand, which is to be avoided where possible. In addition, the dry season is a time when many traditional festivities, such as weddings, family visits and trading in surplus crops are done. These activities are typically planned at this time because there is adequate food from the rain-fed harvest, and also because overall farming activities are at their lowest level, being limited mainly to cultivation in *dimba.* In addition, these traditional festivals are viewed as important occasions and activities for which time and space should be created. Planting on the flat is one such option that makes the job easier by avoiding the need to apply additional water to crops, thereby creating more time for social activities.

Cultivating land that is around the average of three to four hectares per household, with an average family size of five, can be labour demanding. Farmers can become tired and exhausted from cultivation undertaken during the rainy season. In addition, the dry season marks the beginning of the major harvest period. Maize, the major food crop, is dry and ready for harvesting, transportation, processing and storage. This demands the same exhausted labour to harvest and transport the crop to granaries in the villages. *Mundaukulu* are, on average a kilometre away from villages, with a maximum distance of up to 20 kilometres, for example. It is not surprising that farmers try to save on labour by planting crops on the flat in *dimba,* apart from taking advantage of the moisture regime. In addition, many farmers consider part of the dry season as a holiday. They used the phrase '*chihanya ni nyengo yakupumulira,*' which means that the dry season is a resting period, meant for relaxing and celebrating the hard-earned harvests from the labour-demanding activities of the rainy season.

1.3.4 Summary and Conclusions

Farmers are industrious and develop local knowledges that are applied to various environments with which they are faced in their everyday farming practices. These are re-worked to suit the specific environments, although the general principles can be widely applied across the various environments. Soil fertility management is of paramount importance in crop production and it has been perfected to achieve food security. Some of the techniques used to determine and manage soil fertility are comparable to Western scientific knowledges, such that there are complementarities between the two knowledges.

Soil improvement practices in *dimba* are influenced by other activities in the rainy season and during the dry season, and therefore may be given less attention. However, in upland gardens more attention to detail is paid to soil improvement practices, so as to achieve effective and efficient nutrient replenishment, probably because the major food production is carried out in such gardens. Farmers' landholdings in *dimba* are very small by the very nature of wetland size. It does not adequately provide for all household food needs, and hence *dimba* cultivation is seen only as supplementing household food requirements that are largely produced from *mundaukulu*. Farmers also realise the need to live a fuller life by giving themselves time to relax and be on holiday. It appears that development experts have paid very little attention to this fact, and, at times, consider farmers as lazy and unwilling to adopt technologies that can improve their income earnings.

Local knowledges have various uses and roles in society, such as their use in maintaining acceptable socio-cultural values, expectations and norms. This applicability of local knowledge in socio-cultural, economic and various environments makes it hard to sustain the view by some experts that it is backward and static. It further demonstrates that local knowledge is used by people who are well-informed and make rational decisions about their environment, lives and livelihoods, based on 'experimentation' and careful observations that resemble scientific ways of knowing. In addition, the teaching and inter-generational learning in the production of indigenous knowledge challenges the Western notion of the 'expert' and its modernization of agriculture. It is, however, unfortunate that other pressures such as increase in population force farmers in the study area to embark on unsustainable agricultural practices such as opening fields that are not suitable for cultivation even under local ways of cultivation site selection. The correct knowledge on cultivation site selection is available but is knowingly ignored when other forces come into play.

Chapter 2

MODERNIZATION OF AGRICULTURE

2.1 Introduction

Modernization is a theory and practice that highlights the positive role played by the developed world in modernizing and facilitating development (including agriculture) in underdeveloped nations (particularly from 1945 onwards after Truman's bold programme to transfer technology to the 'South').[31] Economic rationality, based on Western intellectual and scientific evolutionism, underpins the development discourse in modernization theory and practice.[32] The theory adopts the original assumption of orthodox development theory, that the process of development involves drawing the populations of the third world out of their traditional isolation into a modern social system that would be participative, pluralistic and

---

[31]C. Leys, *The Rise and Fall of Development Theory*. London: James Currey, 1996; A. Narman, "Getting Towards the Beginning of the End for Traditional Development Aid: Major Trends in Development Thinking and its Practical Application over the Last Fifty Years"; Editors D. Simon and A. Narman, *Development as Theory and Practice*, Harlow: Longman, 1999, pp. 149-180; Z. Sardar, "Development and the Location of Eurocentrism"; Editors R. Munck, and D. O'Hearn, *Critical Development Theory: Contributions to a New Paradigm*, London: Zed Books, 1999.

[32]R. Munck, "Political Programmes and Development: The Transformative Potential of Social Democracy." Editor F.J. Schuurman, *Beyond the Impasse: New Directions in Development Theory*, London: Zed Books, 1993; H. Hove, "Critiquing Sustainable Development: A Meaningful Way of Mediating the Development Impasse?" *Undercurrent vol* 1(1), 2004, pp 48-54. Retrieved on 6th June 2008 from http://forbin.mit.edu/Risk And Preferences/decision models.jsp.

democratic.[33] This is not a new phenomenon, as Buchanan wrote in the context of Nyasaland in 1885:

> "We have entered the Shire highlands for the express purpose of developing the country, and civilizing and Christianizing the natives; and we need not expect that they will be capable for many years to come of attributing to us motives beyond what they themselves ever dreamt of."[34]

He further refers to Africa as a dark continent.[35] Indeed, there are now some scholars who critique and challenge the predominantly North-South flows of knowledge and resources that colonists dedicate human and financial resources towards bringing civilization to primitive backward areas.[36]

However, modernization theory was an American response to what made societies unresponsive to the positive orthodoxy. It emerged from sociologists and political scientists in the United States of America in order to exercise its new role as a super power.[37] They believed that, in the

---

[33]N. Hansen, *Development from Above: The Centre-Down Development Paradigm*. Editors W.B. Stohr, and D.R.F Taylor, Development from Above or Below? The Dialects of Regional Planning in Developing Countries. Chichester, John Wiley and Sons, 1981, pp. 15-38; D.K. Forbes, *The Geography of Underdevelopment*. London: Croomhelm, 1986, p. 57; J. Brohman, *Popular Development: Rethinking the Theory and Practice of Development*. Oxford: Blackwell, 1996, p. 22; C. Leys, *The Rise and Fall of Development Theory*. London: James Currey, 1996 p. 10; R. Munck, "Political Programmes and Development: The Transformative Potential of Social Democracy. Editor F.J. Schuurman, *Beyond the Impasse: New Directions in Development Theory*, London: Zed Books, 1993; R.B. Potter, T. Binns, J.A. Elliot and D. Smith, *Geographies of Development*, Harlow: Pearson Education, 2004, pp. 83, 84.

[34]J. Buchanan, *The Shire Highlands (East and Central Africa) as Colony and Mission*. Edinburgh and London: William Blackwood and Sons, 1885, p. 108.

[35]Ibid p. 2; M. Power, *Rethinking Development Geographies*. London: Routledge, 2003, pp. 76, 139.

[36]C.S. Stringer, C. Twyman and L. Gibbs, "Learning from the South: Common Challenges and Solutions for Small-Scale Farming," *The Geographical Journal* 174 (3), 2008, pp. 235-250.

[37]R. Munck, "Political Programmes and Development: The Transformative Potential of Social Democracy," Editor F.J. Schuurman, *Beyond the Impasse: New Directions*

transition from traditional to modern forms of social organization already completed in the West, the complex interactions between social change and economic development could be traced using structural-functional analysis and a typology of social structures. These were concerned with norms, structure and behaviour in terms of values and interests, plus roles and role networks and classes, not least in terms of motivation and perception. In agriculture, this involved changing the ways of farming from small-scale traditional subsistence geared units dominant in African

in *Development Theory*, London: Zed Books, 1993; Arturo Escobar, *Encountering Development: the Making and Unmaking of the Third World*. Princeton: Princeton University Press, 1995, p. 21; C. Leys, *The Rise and Fall of Development Theory*. London: James Currey, 1996, p. 9; V. Tucker, "The Myth of Development: a Critique of a Eurocentric Discourse," Editors R. Munck and D. O'Hearn, *Critical Development Theory: Contribution to a New Paradigm*," London: Zed Books, 1999; M. Power, *Rethinking Development Geographies*. London: Routledge, 2003, pp. 27, 80.

agriculture to large modern farms that deploy technologically advanced ways of farming that has the sole objective of maximizing production and profits.

In addition to the above, modernization theory highlighted the process of development occurring through diffusion. The concept of 'growth poles' was adapted and re-conceptualized as centres from which technologies would diffuse to hinterlands.[38] For example, development efforts were concentrated in urban areas, which were supposed to diffuse modernization and modern social organization to peripheral areas (Power, 2003:80). Modern agriculture was to provide capital for the development of urban units in addition to enabling the maintenance of low wages through reduced food prices for workers in urban areas. Indeed, Lele, World Bank and Chirwa point out that the government of Malawi's policy to increase food production was considered necessary with specific intention to lower food prices for the growing urban population, for example.[39] Since modernization theory was developed at a time when there were two major political forces at the global scale, the socialist/communist (China and the Soviet Union) and the capitalist (most of the so-called West), democratization became an important element of modernization theory.[40] Aid provision from the West was tied to this democratization process, through the use of aid programmes, technical assistance and provision of

---

[38]N. Hansen, "Development from Above: The Centre-Down Development Paradigm," Editors W.B. Stohr, and D.R.F Taylor, *Development from Above or Below? The Dialects of Regional Planning in Developing Countries*, Chichester: John Wiley and Sons, 1981, pp. 15-38; M. Power, *Rethinking Development Geographies*.Routledge, London: 2003, p.78; R.B. Potter, T. Binns, J.A. Elliot, and D. Smith, *Geographies of Development*, Harlow: Pearson Education, 2004, p. 110.

[39]U. Lele, "Structural Adjustment, Agricultural Development and the Poor, Lessons from the Malawian Experience": The International Bank for Reconstruction and Development/The World Bank, 1989. Retrieved on 28th October 2005 from http://www-wds.worldbank.org; World Bank., "Indigenous Knowledge for Development: a Framework for Action. Knowledge and Learning Centre", Africa Region, World Bank, 1998. Retrieved on 12th April 2007 from http://www.world bank.org/afr/ik, and Chirwa, E.W., "Access to Land, Growth and Poverty Reduction in Malawi Policy Brief," University of Malawi, Chancellor College, Zomba, 2005.

[40]M. Power, *Rethinking Development Geographies*. London: Routledge, 2003, pp. 31, 77.

scholarships to the global South, especially Africa partly to discourage socialist sympathies.[41]

Modernization as a theory therefore focuses on economic growth, despite the fact that it was developed by political and social scientists that sought to incorporate social and political factors in explaining development failures.[42] In agriculture, it is geared towards maximization of profits. Modernization is based on the assumption that backwardness and traditionalism of societies are barriers to progress. Indeed, traditional African farming practices are seen to be backward and inadequate for profit making and subsequent development.[43] Thus modernization defines needs, values and appropriate culture needed for development to take place and relies on the transfer of science and technology from the West to the global South.[44] The problem of poverty and low production in agriculture in developing countries is to be treated by technology transfer, by capital investment and by the release of productive forces. In this context,

---

[41]C. Leys, *The Rise and Fall of Development Theory*. London: James Currey, 1996, p. 10; D.S. Tevera, "Do they Need Ivy in Africa? Ruminations of an African Geographer Trained Abroad," Editors D. Simon, and A. Narmann, *Development as Theory and Practice*, Harlow: Longman, 1999.

[42]V. Tucker, "The Myth of Development: a Critique of a Eurocentric Discourse," Editors R. Munck and D. O'Hearn, *Critical Development Theory: Contribution to a New Paradigm*, London: Zed Books, 1999.

[43]See R. Chambers, *Rural Development: Putting the Last First*, London: Longman, 1983.

[44]R. Munck, "Political Programmes and Development: The Transformative Potential of Social Democracy." Editor F.J. Schuurman, *Beyond the Impasse: New Directions in Development Theory*, London: Zed Books, 1993; Z. Sardar, "Development and the Location of Eurocentrism," Editors R. Munck and D. O'Hearn, *Critical Development Theory: Contributions to a New Paradigm*, London: Zed Books, 1999; V. Tucker, "The Myth of Development: a Critique of a Eurocentric Discourse," Editors R. Munck and D. O'Hearn, *Critical Development Theory: Contribution to a New Paradigm*, London: Zed Books, 1999; M. Power, *Rethinking Development Geographies*. London: Routledge, 2003, p. 75; H. Hove, "Critiquing Sustainable Development: A Meaningful Way of Mediating the Development Impasse?" *Undercurrent vol* 1(1), 2004, pp 48-54. Retrieved on 6th June 2008 from http://forbin.mit.edu/Risk And Preferences/decision models.jsp.

Western scientists speak for the global South.[45] The development agenda is defined in the corridors of power in the 'North' and, in these, the voices of the 'South' are largely unheard.[46]

For modernists, production is a function of the factors of production of land, labour and capital.[47] The bourgeoisie accumulate capital and the proletariat become a factor of production that earns money by working.[48] Restructuring, therefore, is at the expense of social conditions. Production ideas and social relations themselves perpetuate inequality. For the modernist, the important aspects are the types of forces in the production process, such as demand and supply, which influence prices leading to the ultimate goal of profit making or taking.

Modernization approaches therefore envision development even in agriculture as a process of rapidly induced changes that cumulatively result in linear progress toward an end point, closely resembling the contemporary advanced capitalist world.[49] Indeed, postcolonial states have been principally concerned with agricultural modernization, and this has

---

[45]P. Blaikie, "Development, Post-, Anti-, and Populist: a Critical Review," *Environment and Planning* A 32, 2000, pp. 1033-1050; J. Briggs, and J. Sharp, "Indigenous Knowledges and Development: a Postcolonial Caution," *Third World Quarterly* 25 (4), 2004, pp. 661-676; quotes Hooks, 1990 and Spivak, 1988; H. Hove, "Critiquing Sustainable Development: A Meaningful Way of Mediating the Development Impasse?" *Undercurrent vol* 1(1), 2004, pp 48-54. Retrieved on 6th June 2008 from http://forbin.mit.edu/Risk And Preferences/decision models.jsp.

[46]J. Briggs, M. Badri and A.M. Mekki, "Indigenous Knowledges and Vegetation Use among Bedouin in the Eastern Desert of Egypt," *Applied Geography* 19, 1999, pp. 78-103.

[47]M. Power, *Rethinking Development Geographies*. London: Routledge, 2003, p. 30.

[48]H. Hove, "Critiquing Sustainable Development: A Meaningful Way of Mediating the Development Impasse?" *Undercurrent vol* 1(1), 2004, pp 48-54. Retrieved on 6th June 2008 from http://forbin.mit.edu/Risk And Preferences/decision models.jsp.

[49]See J. Brohman, *Popular Development: Rethinking the Theory and Practice of Development*. Oxford: Blackwell, 1996, p. 21; R. Munck, *Political Programmes and Development: The Transformative Potential of Social Democracy*. Editor F.J. Schuurman, *Beyond the Impasse: New Directions in Development Theory*, London: Zed Books, 1993.

translated itself into a desire to plan, at rapid pace, the profound social transformation of the countryside.[50] Harrison demonstrates that such transformation is defined by Western scientific methods, attempts to increase output in cash crops, and it rarely involves meaningful participation by the peasantry.[51] For modernization theorists, the middle and the upper classes are seen to be crucial for this process. Progressive members of these classes play the role of a modernizing elite, as change agents who bear the modern entrepreneurial values, hence the provision of scholarships.[52] These change agents are supposed to be involved in the process of democratization and development of agriculture. However, this process is silent on bottom-up approaches, which are naturally associated with the process of democracy and agricultural development.[53] For example, Briggs note that 'experts' consider that no worthwhile contribution can be made by the inhabitants of fragile environments, as they have little meaningful to offer; indeed, left to their own ways will only result in further degradation.[54] For many in the 'North', lands have to be managed in a rational technocratic manner using knowledge rooted firmly in formal Western science and technology.[55]

---

[50]G. Harrison, "Peasants, the Agrarian Question and Lenses of Development," *Progress in Development Studies* 1 (3), 2001, pp. 187-203.

[51]Ibid.

[52]J. Brohman, *Popular Development: Rethinking the Theory and Practice of Development*. Oxford: Blackwell, 1996, p. 22; C. Leys, *The Rise and Fall of Development Theory*. London: James Currey, 1996, p. 10; A. Narman, "Getting Towards the Beginning of the End for Traditional Development Aid: Major Trends in Development Thinking and its Practical Application over the Last Fifty Years," Editors D. Simon and A. Narman, *Development as Theory and Practice*, Harlow: Longman, 1999, pp. 149-180; D.S. Tevera, "Do they Need Ivy in Africa? Ruminations of an African Geographer Trained Abroad," Editors D. Simon, and A. Narmann, *Development as Theory and Practice*, Harlow: Longman, 1999.

[53]M. Power, *Rethinking Development Geographies*, London: Routledge, 2003, p. 81.

[54]J. Briggs, M. Badri and Mekki, A. M., "Indigenous Knowledges and Vegetation Use among Bedouin in the Eastern Desert of Egypt," *Applied Geography* 19, 1999, pp. 78-103.

[55]Ibid; P. Blaikie, "Development, Post-, Anti-, and Populist: a Critical Review," *Environment and Planning,* A 32, 2000, pp. 1033-1050.

The common indices used to measure progress along the road to development are factors thought to be responsible for growth among modern Western societies, such as, for example, the growth of formal rationality and of complex bureaucratic organizations, increased social differentiation and industrial (including agriculture) specialization.[56] Yet, the area where this modernization process is to take place, especially Africa, still has a sense of community responsibility and individual rationality is influenced by communal values, as can be found in communal land ownership and agricultural production objectives.[57] Furthermore, vernacular societies have developed ways of defining and treating poverty that accommodates visions of community, frugality, and sufficiency.[58]

Modernization theorists envisage modern values being diffused through formal education and technology transfer to the elites of the periphery.[59] In this context, the values, interests and norms to be promoted are those of Western culture. All developing countries need to do is to emulate the most successful societies and cultures for them to move out of their 'traditional isolation' into a modern social system that include modern ways of farming.[60] The villagization which was a process of creating communal

---

[56]J. Brohman, *Popular Development: Rethinking the Theory and Practice of Development*, Oxford: Blackwell, 1996, p. 21; S. Morse, "The Geography of Tyranny and Despair: Development Indicators and the Hypothesis of Genetic Inevitability of National Inequality," *The Geographical Journal* 174 (3), 2008, pp. 195-206.

[57]See B.H.Z. Moyo, "The Use and Role of Indigenous Knowledge in Small-scale Agricultural Systems in Africa: the Case of Farmers in Northern Malawi," Published PhD thesis. Glasgow University, 2008. http://theses.gla.ac.uk

[58]Arturo Escobar , *Encountering Development: the Making and Unmaking of the Third World*, Princeton University Press, 1995, p. 22.

[59]J. Brohman, *Popular Development: Rethinking the Theory and Practice of Development*. Oxford: Blackwell, 1996, p. 22; C. Leys, *The Rise and Fall of Development Theory*. London: James Currey, 1996, p. 10; M. Power, *Rethinking Development Geographies*. London: Routledge, 2003, p. 80.

[60]See G. Harrison, "Peasants, the Agrarian Question and Lenses of Development," *Progress in Development Studies* 1 (3), 2001, pp. 187-203; Arturo Escobar , *Encountering Development: the Making and Unmaking of the Third World*. Princeton University Press, 1995, p. 26; J. Brohman, *Popular Development: Rethinking the Theory and Practice of Development*. Oxford: Blackwell, 1996, pp.

villages in Mozambique and Tanzania so that agricultural activities and other social amenities could be efficiently managed and provided respectively by the governments concerned (its major advantage), is an example of the imposition of modern social system on rural small-scale farmers in Africa.[61] In reality, it means relegating indigenous values to the bottom of the spectrum of values, as progressive members of the elite classes play the role of change agents by becoming the bearers of modern values, especially those of entrepreneurship.[62] This contrasts with the awareness of the existence of complex interactions between social change and economic development of social and political scientists. Calls are now being made to recognize that such cultural values and knowledges may be important in the development of modernization theory itself and probably modernization of agriculture too.[63]

19, 20; R. Munck, *Political Programmes and Development: The Transformative Potential of Social Democracy*. Editor F.J. Schuurman, *"Beyond the Impasse: New Directions in Development Theory,"* London: Zed Books, 1993; M. Power, *Rethinking Development Geographies*. London: Routledge, 2003, p. 79; R.B. Potter, T. Binns, J.A. Elliot and D. Smith, "Geographies of Development," Harlow: Pearson Education, 2004, p. 84; D. Simon, "Beyond Antidevelopment: Discourses, Convergence, Practices," *Singapore Journal of Tropical Geography* 28, 2007, pp. 205-218.

[61]See G. Harrison, "Peasants, the Agrarian Question and Lenses of Development," *Progress in Development Studies* 1 (3), 2001, pp. 187-203.

[62]V. Tucker, "The Myth of Development: a Critique of a Eurocentric Discourse," Editors R. Munck and D. O'Hearn, *Critical Development Theory: Contribution to a New Paradigm*, London: Zed Books, 1999; M. Power, *Rethinking Development Geographies*. London: Routledge, 2003, p. 76; C. Eriksen, "Why do they Burn the 'Bush'? Fire, Rural Livelihoods, and Conservation in Zambia," *The Geographical Journal* 173 (3), 2007, pp. 242-256; J.A. Riseth, "An Indigenous Perspective on National Parks and Sami Reindeer Management in Norway," *Geographical Research* 45 (2), 2007, pp. 177-185; J. Brohman, *Popular Development: Rethinking the Theory and Practice of Development*. Oxford: Blackwell, 1996, p. 22.

[63]R. Chambers, *Rural Development: Putting the Last First*. London: Longman, 1983, R. Chambers, *Challenging the Professions: Frontiers for Rural Development*. London: Intermediate Technology Publications, 1993; A. Bebbington, "Modernization from Below: an Alternative Indigenous Development?" *Economic Geography* 69, 1993, pp. 274-292; P. Blaikie, "Development, Post-, Anti-, and Populist: a Critical Review,"

In the development paradigm, former colonies are to adopt Western norms and values so that development will be induced through a top-down, centre-outward process of capitalist development via modernizing elites. The elite are sent to universities in the West to become leaders in the development process and trained in Western science and technology, which is considered superior to traditional local knowledges.[64] Indeed, Eriksen notes that post-colonial leaders were educated in rigid colonial government systems including agricultural management methods so that many colonial policies were retained as newly independent states pursued economic development.[65] In effect, this translated into practice, so that the norms, values, interests, roles, role networks and classes that existed in the colonies and former colonies were inadequately utilized on the basis that Western culture and its agricultural production methods were superior to traditional social organization and agricultural production systems.[66] The result is that some development projects introduced by development experts have been inappropriate and even irrelevant.[67]

Environment and Planning A 32, 2000, pp. 1033-1050; J. Briggs, and J. Sharp, "Indigenous Knowledges and Development: a Postcolonial Caution," *Third World Quarterly* 25 (4), 2004, pp. 661-676; C. Eriksen, "Why do they Burn the 'Bush'? Fire, Rural Livelihoods, and Conservation in Zambia," *The Geographical Journal* 173 (3), 2007, pp. 242-256.

[64]R. Chambers, *Rural Development: Putting the Last First*, London: Longman, 1983; R. Chambers, *Challenging the Professions: Frontiers for Rural Development*. London: Intermediate Technology Publications, 1993.

[65]C. Eriksen, "Why do they Burn the 'Bush'? Fire, Rural Livelihoods, and Conservation in Zambia," *The Geographical Journal* 173 (3), 2007, pp. 242-256.

[66]See R. Munck, *Political Programmes and Development: The Transformative Potential of Social Democracy*. Editor F.J. Schuurman, *Beyond the Impasse: New Directions in Development Theory*, London: Zed Books, 1993; J.N. Pieterse, "Critical Holism and the Tao of Development" Editors R. Munck and D. O'Hearn, *Critical Development Theory: Contribution to a New Paradigm*, London: Zed Books, 1999, pp. 63-88; Z. Sardar, "Development and the Location of Eurocentrism," Editors R. Munck and D. O'Hearn, *Critical Development Theory: Contributions to a New Paradigm*, London: Zed Books, 1999; M. Power, *Rethinking Development Geographies*, London: Routledge, 2003, p. 75.

[67]R. Chambers, *Rural Development: Putting the Last First*. London: Longman, 1983.

Development could, therefore, be conceived as the training of role models, resulting in a smooth transition from traditional societies to modern forms of organization. However, the transition is often bumpy. It became clear that the process of political change accompanying modernization in many Third World countries in the global South has been far from smooth since the 1960s. It has become disruptive and at times serves as a source of political conflict.[68]Peters notes that the compulsion in promoting soil-conservation techniques in Malawi became more marked by the 1940s and were one of the sources of resentment among rural people that fed into the political resistance of the 1950s that eventually led to independence in 1964.[69] Development theorists are now urged to pay more attention to cultural and political structures and to incorporate both objective and subjective elements into their analysis.[70]

The process is facilitated by the provision of development experts who analyze the problems and offer technical and scientific solutions to development problems *in situ*.[71] The questions that arise over time are whether the training is adequate for the conditions on the ground; are experts knowledgeable about local conditions; and whether the projects borne out of such assessment by development experts reflect local needs and ambitions.[72]

The transfer of science and technology is one of the most important aspects of modernization theory.[73] It aims at dealing with poverty and under production of resources (land, labour and capital) through the

[68]J. Brohman, *Popular Development: Rethinking the Theory and Practice of Development*. Oxford: Blackwell, 1996, p. 22.

[69]P.E. Peters, "The Limits of Knowledge: Securing Rural Livelihoods in a Situation of Resource Scarcity," Editors C.B. Barrett, F. Place and A.A. Aboud, *Natural Resources Management in African Agriculture Understanding and Improving Current Practices*, Oxon, CABI Publishing, 2002.

[70]J. Brohman, *Popular Development: Rethinking the Theory and Practice of Development*. Oxford: Blackwell, 1996, p. 24.

[71]P. Blaikie, "Development, Post-, Anti-, and Populist: a Critical Review," *Environment and Planning* A 32, 2000, pp. 1033-1050; M. Power, *Rethinking Development Geographies*, London: Routledge, 2003, p. 75.

[72]J. Townsend, "Gender Studies: Whose Agenda?" Editor F.J. Schuurman, *Beyond the Impasse: New Directions in Development Theory*, London: Zed Books, 1993.

transfer of science and technology.[74] What has worked in the development of the Western economies is to be reproduced in the former colonies in Africa and elsewhere.[75] Development and agricultural production assumes a linear relationship in terms of capital, land and labour and the manipulation of these using science and technological advances. Essentially, modernization theory assumes that societies are made up of rational individuals, collectively making their way through evolution, heading in the same direction, and that differences between societies can be understood primarily as differences in development.[76] The approach does little to eradicate poverty and nothing to resolve the degradation of the environment.[77]

---

[73]Z. Sardar, "Development and the Location of Eurocentrism," Editors R. Munck and D. O'Hearn, *Critical Development Theory: Contributions to a New Paradigm*, London: Zed Books, 1999.

[74]J. Briggs, M. Badri and A.M. Mekki, "Indigenous Knowledges and Vegetation Use among Bedouin in the Eastern Desert of Egypt," *Applied Geography* 19, 1999, pp. 78-103.; J. Briggs and J. Sharp, "Indigenous Knowledges and Development: a Postcolonial Caution," *Third World Quarterly* 25 (4), 2004, pp. 661-676.

[75]F.J. Schuurman, "Development Theory in the 1990s," Editor F.J. Schuurman, *Beyond the Impasse: New Directions in Development Theory*," London: Zed Books, 1993; J. Brohman, *Popular Development: Rethinking the Theory and Practice of Development*. Oxford: Blackwell, 1996, p. 24; Z. Sardar, "Development and the Location of Eurocentrism," Editors R. Munck and D. O'Hearn, "Critical Development Theory: Contributions to a New Paradigm," London: Zed Books, 1999.

[76]V. Tucker, "The Myth of Development: a Critique of a Eurocentric Discourse," Editors R. Munck and D. O'Hearn, "Critical Development Theory: Contribution to a New Paradigm," London: Zed Books, 1999; M. Power, *Rethinking Development Geographies*. London: Routledge, 2003, p. 75; H. Hove, "Critiquing Sustainable Development: A Meaningful Way of Mediating the Development Impasse?" *Undercurrent vol* 1(1), 2004, pp 48-54. Retrieved on 6th June 2008 from http://forbin.mit.edu/Risk And Preferences/decision models.jsp.

[77]H. Hove, "Critiquing Sustainable Development: A Meaningful Way of Mediating the Development Impasse?" *Undercurrent vol* 1(1), 2004, pp 48-54. Retrieved on 6th June 2008 from http://forbin.mit.edu/Risk_and_Preferences/decision models.jsp.

2.2 Modern Agriculture

Modernization of agriculture involves the application and transfer of science and technology from experts to farmers (uni-directional) as understood in the developed world in pursuit of increased production and productivity.[78] Increased production involves expansion of area under cultivation or used to rear livestock while increased productivity refers to increasing production per unit input that includes land, labour and capital. The emphasis has been specialization by maximizing production of one commodity that makes the most profit than increasing total farm output from diversification. In crop production, plants which compete with the desired crops for water, sunlight, space and nutrients are regarded as weeds, for example, even if they can be of use for consumption and/or sale. High yielding varieties such as hybrids are promoted. Such crops can be genetically modified or produced through the traditional breeding programmes. The cultivation practices involves clean (weed free) seedbed throughout the crop life in the garden through weeding or application of herbicides, optimum use of inorganic fertilizers and the application of pesticides to controls pests and diseases. Most of such management practices (cultivation, weeding and application of fertilizer and pesticides) are accomplished through the use of machinery including tractors. Modernization of agriculture has concentrated on reducing the competition crops face while growing in the fields.

Modernization represents the achievements of advanced, industrial societies and defines the objectives of peasants and non-industrial societies as denoting cultural and technologically backwardness.[79] Local practices in Malawi such as those that include leaving weeds in some crops as a nutrient recycling and soil protection mechanism as shown in figure 1 thus denotes society's ignorance and backwardness.[80]

---

[78]G. Harrison, "Peasants, the Agrarian Question and Lenses of Development," *Progress in Development Studies* 1 (3), 2001, pp. 187-203.

[79]Ibid.

[80]See also B.H.Z. Moyo, "The Use and Role of Indigenous Knowledge in Small-scale Agricultural Systems in Africa: the Case of Farmers in Northern Malawi," Published PhD thesis. Glasgow University, 2008. http://theses.gla.ac.uk

Figure 2.1 Weeds left in a maize garden

It is therefore ironical that some modern technologies such as minimum tillage subscribe to the local practices of leaving weeds as a soil erosion protection mechanism.

In livestock increased off take (production) is mainly achieved through replacement of local with exotic breeds although improving local breeds through cross-breeding is sometimes advocated where resistance to pests and diseases from local breeds is the major objective.[81] In addition, where local farmers are considered 'inadequately' trained in 'good' livestock management practices, the provision of crossbreds acts as a training mechanism for graduation into the successful management of exotic pure breeds. In this modernization process unfortunately, locally available pastures are considered unsuitable for exotic breeds and improved

---

[81]See S. Kratli, "What do Breeders Breed? On Pastoralists, Cattle and Unpredictability," *Journal of Agriculture and Environment for International Development* 102 (1-2), 2008, pp. 123-139; B.H.Z. Moyo, "The Use and Role of Indigenous Knowledge in Small-scale Agricultural Systems in Africa: the Case of Farmers in Northern Malawi," Published PhD thesis. Glasgow University, 2008. http://theses.gla.ac.uk

pastures in form of exotic plants are brought in to maintain high production levels of the introduced exotic breeds in meat or milk yields.[82]

The increased production as a result of increased allocation of land to crops or livestock and the need to satisfy market requirements have associated cultural norms and practice change that can be problematic. The local ownership of land is often considered insecure under market oriented production 'as it is seen to' fail to enable land to become collateral for acquiring capital from institutions such as commercial banks. Land ownership is then converted from customary to a form such as leasehold tenure which is recognized by Western derived institutions including courts although such ownership of the land is alien to local traditions. Unfortunately, the change of land tenure to forms like leasehold and free-hold that are recognized by courts and commercial banks deprive local people of their title to land and subsequently increases land ownership to commercial-scale farmers who can adequately pay land rents. At individual farmer level, loss of land means loss of security and the means of production to sustain ones life as land is the single most important factor of production so that the loss of its (land) ownership is the surest way of local farmers' movement into absolute poverty and deprivation. Land moves from subsistence type of production that has been perfected for a long time under local conditions to commercial production that has a lot of associated negative environmental and socio-cultural related problems.

There is an increased production of crops under mono-cropping system which puts pressure on the environment. The large area of production under mono-cropping provides an excellent breeding ground for pests and diseases as their food is available in abundance to sustain large populations of pests and disease causing organisms. Of paramount importance is the potential loss of indigenous farming practices that may arise from loss of land on which to cultivate. Thompson observes that inherent in indigenous knowledge transfer is its inability to skip generations.[83]Thompson's findings

---

[82]See B.H.Z. Moyo, "The Use and Role of Indigenous Knowledge in Small-scale Agricultural Systems in Africa: the Case of Farmers in Northern Malawi," Published PhD thesis. Glasgow University, 2008. http://theses.gla.ac.uk

[83]I.B. Thompson, "The Role of Artisan Technology and Indigenous Knowledge Transfer in the Survival of a Classic Cultural Landscape: the Marais Salants of Guerande, Loire-Atlantique, France," *Journal of Historical Geography* 25 (2), 1999, pp. 216-234.

establish that if local means of production is not passed on to the next generations it is likely to be lost. It appears the consequences of modernization of agriculture if not carefully thought through are many.

2.3 Some Side Effects of Modernization of Agriculture

2.3.1 Pesticides

As noted above, the increased areas under cultivation particularly under mono-cropping and intensive (confined) livestock production such as chickens reared in battery cages increase the chances of incidences of pests and diseases. This necessitates the application of pesticides that have many negative environmental effects and impacts. Pesticides are by nature harmful to organisms they are meant to control. However, the harmful effects are not limited to the target organisms. Pesticides do kill unintended useful organisms such as earthworms that improve soil aeration and structure; and bees useful in pollination of crops and wild plants. Many other scholars have demonstrated that pesticides are also harmful to human beings showing that, yearly, thousands of people are poisoned by pesticides. Furthermore, overtime pests and disease causing organisms build up resistance to pesticides, which must then be used in even higher doses to have the desirable effects. Eventually new pesticides must be developed, which increases the diversity of harmful products being introduced into the environment. If such products do not breakdown easily, they then accumulate in the environment including in the food chain or food webs causing considerable damage to insects, insect consuming animals, prey birds and ultimately human beings.

2.3.2 Impacts of Cultivation

Although, the soil is viewed as a fragile and living medium that must be protected and natured to ensure its long-term productivity and stability especially by many smallholder farmers, large scale production demands the use of mechanization such as the utilization of tractors for cultivation.[84] The seed bed is cleared free of weeds through soil pulverization using implements such ploughs drawn by tractors. The creation of a suitable seedbed for crop cultivation results in soil loosening, compaction and consolidation.

This is called tillage. Tillage thus is a process through which soil condition is normally improved for the cultivation of crops. However, tillage affects soil structure, water holding capacity, aeration, infiltration capacity, temperature and evapo-transpiration. Furthermore, tillage (tilling the soil) has some negative impacts that include the breaking of capillary connections in the soil, the quick drying of the top layer of soil apart from loosening it so that the soil becomes prone to erosion by erosive agents such as run-off and wind. Further still, tillage increases the mineralization of organic matter resulting in loss of this important soil component. Yet a 'healthy' soil is a key component of sustainability; that is, a healthy soil will produce crop plants that have optimum vigour and are less susceptible to pests and diseases.[85] Indeed, crop management systems such as soil pulverization that impair soil quality often result in greater inputs of water, nutrients, pesticides, and/or energy for tillage to maintain yields. Of significance is the tillage requirement of energy. Tractors, oxen and manpower are all sources of energy required to till the land. Tractors use fossil fuels (petrochemicals) that are non-renewable and are a source of gases (such as carbon dioxide) that have a negative impact on climate (global warming) apart from directly polluting the environment. Oxen are known to produce methane that has a negative impact on climate. Fortunately, oxen have the advantage of using renewable resources for their energy generation. Dung, a waste product from food consumed by oxen can improve the soil condition to which it is applied. Manpower is needed to run tractors as well as operate ox-drawn implements. However, it can be utilized directly in cultivation of crops through the use of a hoe. Manpower can have negative impacts on the environment when conspicuous consumption is adopted. This creates excessive use of energy

---

[84]See B.H.Z. Moyo, "The Use and Role of Indigenous Knowledge in Small-scale Agricultural Systems in Africa: the Case of Farmers in Northern Malawi," Published PhD thesis. Glasgow University, 2008. http://theses.gla.ac.uk

[85]University of California, "What is Sustainable Agriculture?" http://www.sarep. ucdavis.edu, 2010.

and resources resulting in a lot of waste finding its way to landfills or dumping sites.

2.3.3 Impacts Associated with Livestock Production

As we have seen above, livestock waste including dung is useful and can be used as a source of nutrients such as nitrogen, phosphorus and potassium for crops in form of manure. Manure not only improves soil nutrient content but also improves the soil water holding capacity and soil structure that is so important in the control of soil erosion and the increase in crop yields.

However, leaching and volatization can result in nutrients found in animal wastes used as manure finding their way into water bodies. When nutrients such as nitrogen and phosphorus contaminate water bodies, they result is water pollution that is known as eutrophication. Where there is a slaughter house (abattoir) next to a stream, the wastes are normally disposed into the stream. Such practices result in point source pollution.

The consequent excessive nutrients from point sources and non-point sources when they find their way in to water bodies encourage aquatic plant growth such as algae. When such plants (algae) die, they cause water (colour) discolouration, increased chemical oxygen demand (COD), increased biological oxygen demand (BOD) and changes in water pH. All these changes may affect negatively aquatic life such as the type of fish that would thrive in such waters. Aquatic lives that require oxygen for respiration may be replaced by those that use anaerobic conditions to generate energy necessary for life sustenance.

As we have seen, livestock such as cattle and sheep may require improved pastures to maintain high meat and milk yields especially if they are exotic breeds. As mentioned earlier on, this may require introduction of exotic plants. Such introduction of exotic plants translates into change into the ecology of the area to which these new plants are introduced. Biodiversity is then threatened if such pastures become invasive. It is clear that livestock farming, therefore, increases the complexity of biological and economic relationships.

2.3.4 Impacts of Aquaculture on the Environment

Aquaculture is a growing industry as natural sources of fish in rivers, lakes; seas and ocean are being depleted as a result of poor management (of such resources). Scientific knowledge has provided an estimate of the amount of harvest that can result in sustainable utilization of such resources including fish (see European fishing quota system). However, the implementation of a programme to harvest resources in a sustainable manner has proved difficult at both local as well global scales. For example, Malawi has introduced a closed fishing season when fish are breeding normally at the beginning of the rainy season. Such a practice has been difficult to implement and let alone to enforce. Those fishermen borne and living along the lakes and rivers of Malawi are used to fishing all year round and they have maintained their practice despite the introduction of a closed season. The result is continued depletion of fish resources that requires rearing fish in some form – aquaculture is thus the alternative.

Aquaculture may involve the creation of ponds where fresh water is abundant or may be in form of cages in big water bodies such as lakes (see Maldeco fisheries in Mangochi). In most instances where there is need for pond creation, the habitats are modified. In case of forests and river line vegetation the first task is to clear these to make room for operations that involve digging and shaping of ponds to required depths. All these processes change the original nature of the habitat, and the soil is disturbed through being loosened and becomes prone to erosion by wind and run-off.

The introduction of fish in ponds or cages requires that production is maintained at a level where it is profitable. Profit levels are attained when costs of investments are recovered and operational costs can be paid from sales so that a surplus is left that can then contribute to income for the investor. The generation of surplus demands management practices such as those necessary for disease and pest control including proper storage of fish and its onward transportation to markets.

As a consequence intensive fish production in ponds or cages requires feeding them. Unfortunately not all feed given to fish is consumed.[86] Some of it becomes waste and the nutrients contained in such feed become a source of eutrophication. In addition, the feed together with fish excreta

---

[86]See J.G. Jones, "Pollution from Fish Farms" in *Agriculture and Environment* edited by Jones, J.G. Ellis, New York, Horwood. 1993.

change the chemical and biological oxygen demand.[87] The water too is discoloured if the waste is not continuously flushed out of the ponds. When such wastes are flushed out without treatment, the wastes and the effects are simply transferred to the receiving bodies. This likely transfer of pollutants from one medium to the other or from one area to the other calls for proper management of pollutants from fish production.

2.3.5 Farmers' Goals and Lifestyle Choices

Adoption of some technologies or practices that promise profitability may also require such intensive management that one's lifestyle actually changes for the better or worse. For example, the adoption of hybrid varieties may demand use of pesticides that endanger ones life so that it becomes necessary to changes one's lifestyle. Indeed, care must be taken in keeping pesticides away from children, for example. The introduction of pesticides in a family may bring worries for the safety of children. Hybrids may require specialization and mono-cropping that may replace crop diversification, thus endangering that nutritional security obtained from growing different diversified crops at household level apart from the risk of losing steady income and assured food security.

2.4 Summary

The principle behind modernization of agriculture is now becoming difficult to sustain and defend because of numerous side-effects experienced when it is practised. The theory of increasing production to keep on feeding the increasing population and profit maximization may have worked well before the dangers associated with modernization were realized. The gross undermining of the traditional agricultural practices that protect the environment through re-working the knowledge repertoire is now gaining ground. It is thus difficult to ignore it in favour of modern agricultural practices despite their touted benefits. Clearly, there are calls for modernization of agriculture that learns from local agricultural practices that show considerable knowledge of, and sympathy with, the environment.

---

[87]Ibid.

Chapter 3

GLOBALIZATION

3.1 Introduction

The need for increased production to generate profit and feed the growing population of the world has contributed to the creation of the term globalization. The phenomenon associated with globalization is not necessarily new. The colonization of continents such as America and Africa brought about trade that linked these continents creating a global scale production and trade. In Nyasaland now Malawi Buchanan notes the production of crops such as tobacco and sugar cane for export to Europe and in particular to Great Britain.[88] Even slave trade before the settlement of whites in Malawi was at global scale. However, the speed at which information and trade transactions take place today is far much faster than in the past (during slavery and colonization). Overtime, associated with the speed of at which information and trade transactions take place, there has been a transformation of agricultural production and distribution that has increasingly favoured Transnational Companies in 'upstream' and 'downstream' activities from farms.[89]

3.2 The Nature of Globalization

Potter notes that globalization is customarily recognized as consisting of principal strands that include the economic, the cultural and the political[90]. For Potter economic globalization refers to the fact that distance has become less important to economic activities so that large corporations

---

[88]J. Buchanan, *The Shire Highlands (East and Central Africa) as Colony and Mission*. Edinburgh and London: William Blackwood and Sons, 1885.

[89]See R.B. Potter, T. Binns, J.A. Elliot and D. Smith, *Geographies of Development*, Harlow: Pearson Education, 2004.

[90]Ibid.

subcontract to branch organizations in far distant regions, effectively operating within a borderless world; cultural globalization suggests that western forms of consumption and life styles have spread across the world apart from the fact that there is an increasing convergence of cultural styles; and political globalization refers to the internationalization that has lead to the erosion of the former role and powers of the nation-state.[91] This is particularly true with developing countries where multinational companies seem to dictate the nature of relationship between them and the developing nations. In the name of foreign direct investments, the terms of trade and investment are skewed in favour of the needs of foreign investors. For example, Malawi Government received a 15% free carried equity in the Kayelekera uranium mining project in Karonga even though the mineral, uranium, is the property of the people of Malawi and the government.[92]

As depicted by Potter for Malawi the Kayelekera operations are controlled by investors whose base is largely in Australia.[93] Furthermore, the operations are done by experts from all over the world with a few Malawian experts. However, Malawians form the bulk of unskilled labour for which there are low returns on their input. There is also minimum addition of value to the mineral before it leaves the country. The uranium is only partially treated in Malawi with further processing and enrichment being done outside Malawi where labour earns significant return on their input and the parent companies add value to the mineral for higher earnings. Clearly, the case of Kayelekera uranium mine depicts what Potter terms as some common interpretations of the term globalization that includes[94]:

- The worldwide spread of modern technologies of industrial production and communication regardless of frontiers

- The networking of virtually all the world's economies representing a move in the direction of functional integration

---

[91]Ibid.

[92]See http://www.mining-technology.com/projects/kayelekerauraniummin/.
[93]R.B. Potter, T. Binns, J.A. Elliot and D. Smith, *Geographies of Development*, Harlow: Pearson Education, 2004.
[94]Ibid.

- The linking of and inter-relationships between cultural forms and practices that occur when societies become integrated into and dependent on world markets (this is clearly demonstrated also by tobacco industry in Malawi as well)

- The convergence and homogenization of capitalist economic forms and relations across nations {albeit in nature of superiors (developed countries) versus inferiors (developing countries)}.

- And I add that globalization facilitates the continued exploitation of resources of the developing countries by developed countries so that the gap widens in terms of economic growth between these two.

Indeed, we live in a world in which events happening at a particular location are shaped by developments occurring thousands of kilometres away. As noted above, tobacco production and prices in Malawi are shaped by smokers and entrepreneurs elsewhere in the world.[95] Movements of capital, money, information and cultural exchange occur because largely there remain substantial differences between places, regions, and country.[96] It is important to note that not everyone is affected in the same way by the availability of global products, but there is a greater interconnectedness in terms of product availability and consumption. Globalization consequently gives rise to:[97]

- An increase in the power of finance over production. There is a general belief by some scholars that capital can move seamlessly and very rapidly around the world with impacts on regional and national economies that change fortunes of companies and individuals. Interestingly, in Malawi small scale tobacco farmers have become dominant producers replacing the state sector as a result of fluctuating tobacco prices associated with the dominance of multinationals in the industry. As we have seen, the Kayelekera mine demonstrates the power of finance over production. The dominant factor then is that those production systems that are not dependent

---

[95]Malawi Nation Newspaper., "Tobacco Business: who is Duping the Poor Farmer?" 2008. Retrieved on 15th May 2008 from http://www.nationmw.net.

[96]R.B. Potter, T. Binns, J.A. Elliot and D. Smith, *Geographies of Development*, Harlow: Pearson Education, 2004.

[97]Ibid.

on finance are deliberately rated as unsuitable. I note that subsistence production systems that are less dependent on finances are looked upon as backward and responsible for underdevelopment in countries they are still practiced.[98] Interestingly, the multinationals search for large profits for their shareholders by moving capital to where it gets the best returns on the investment. This is usually in developing countries where the cost of production and doing business is low hence outsourcing. Surprisingly, in Malawi for sometime the externally controlled tobacco prices have been at levels that 'favour'[99] the small scale producers who use less finances and external inputs in comparison to estate sectors.

- The pivotal role of knowledge as a vital factor of production (as though this has changed over time). There is a bias towards the belief that there is a significant economic benefit in producing an educated and skilled workforce.[100] Such a belief unfortunately only depicts the importance of formal education at the expense of local 'knowledges' that have been the foundation of progress and development even in developed countries.[101] Indeed, Briggs observes that Western science

---

[98]B.H.Z. Moyo, "The Use and Role of Indigenous Knowledge in Small-scale Agricultural Systems in Africa: the Case of Farmers in Northern Malawi," Published PhD thesis. Glasgow University, 2008.http://theses.gla.ac.uk.; B.H.Z. Moyo, "Indigenous Knowledge-Based Farming Practices: A Setting for the Contestation of Modernity, Development and Progress," *Scottish Geographical Journal* 125 (3-4), 2009, pp. 353-360.

[99]Favour here refers to the fact that small scale farmers have lower external production costs as a result of self-sufficiency (subsistence) in comparison to the estate sector so that profit is realized at reduced tobacco prices although they would be happier with higher prices that would also make the estate sector make profits.

[100]See D.S. Tevera, "Do they Need Ivy in Africa? Ruminations of an African Geographer Trained Abroad," Editors D. Simon, and A. Narmann, "Development as Theory and Practice," Harlow, Longman, 1999.

[101]See J. Briggs, "The Use of Indigenous Knowledge in Development: Problems and Challenges," *Progress in Development Studies* 5 (2), 2005, pp. 99-114; B.H.Z. Moyo, "The Use and Role of Indigenous Knowledge in Small-scale Agricultural Systems in Africa: the Case of Farmers in Northern Malawi," Published PhD thesis. Glasgow University, 2008.http://theses.gla.ac.uk; B.H.Z. Moyo, "Indigenous Knowledge-

is as much socially constructed as indigenous knowledge in addressing the daily demands of their respective societies.[102]

• The global nature of technologies, transportation, communication and mobility of finance and capital has generated domination of production and distribution systems by multinationals. Furthermore, the market of certain products such as tobacco in Malawi is manipulated by large companies such as British American Tobacco (BAT) and Philips and Morris which are the major final buyers of tobacco before it is processed for sell to consumers.

• The result in Malawi is the loss of ability of the nation to regulate its own economic development. This is further complicated and reinforced by the influence of the IMF and the World Bank policies that have been detrimental to development efforts of some countries including Malawi.[103] It is now clear that structural adjustment programmes initiated in the 1980s by the IMF have done more harm than good to the development efforts of Malawi, for example.[104] Currently, the devaluation and the floatation of the Kwacha in 2012 seem to be doing more harm than good to the ordinary Malawian with interest rates hovering around 40% in commercial banks and inflation rising to double digits.

• Globalization has increased the transformation of agricultural products from items of immediate consumption as was the case with tobacco under subsistence production in Malawi into inputs for the greater manufacturing systems as in the production of cigarettes by BAT and Philip and Morris companies, for example. The result has

Based Farming Practices: A Setting for the Contestation of Modernity, Development and Progress," *Scottish Geographical Journal* 125 (3-4), 2009, pp. 353-360.

[102]J. Briggs, "The Use of Indigenous Knowledge in Development: Problems and Challenges," *Progress in Development Studies* 5 (2), 2005, pp. 99-114.

[103]See U. Lele, "Structural Adjustment, Agricultural Development and the Poor, Lessons from the Malawian Experience": The International Bank for Reconstruction and Development/The World Bank, 1989. Retrieved on 28th October 2005 from http://www-wds.worldbank.org.

[104]Ibid; B.H.Z. Moyo, "The Use and Role of Indigenous Knowledge in Small-scale Agricultural Systems in Africa: the Case of Farmers in Northern Malawi," Published PhD thesis. Glasgow University, 2008. http://theses.gla.ac.uk

been the maintenance of developing countries as major producers of 'low value' primary products such as minerals and agricultural produce with developed countries adding value to these and retaining the production of 'high value' products. At the end of the day developing countries become net importers of their (own) produce in the false name of high value products from developed countries (cigarettes being of higher value than the tobacco exported by developing countries so too diamond being of lower value than a necklace made from the same by developed countries).

Globalization thus implies that competition is expressed through the ability of multinationals in regulating prices to a level that reduces the costs, increasing the control over sources of raw materials, transport and distribution systems and through global sourcing.[105] The effects of globalization as we have seen in the case of small scale tobacco production in Malawi have a mixed set of results. In Malawi, it has favoured production systems that are 'less intensive' (so to speak) where costs are spread over several activities such as other crops, livestock and entrepreneurship, which reflect diversified production systems as opposed to specialization. Even where intensive production is depicted as in small holder sugar cane production in Malawi, for example, farmers here benefit from their subsistence farming as a measure of spreading costs over several activities. Of particular importance is where small-scale farmers are still engaged in other activities such as producing crops and livestock that are for household consumption so that cash crops become a form of an extension beyond household needs.[106] However, elsewhere such as in South America, banana production has favoured intensification that has almost eliminated small-scale producers from partaking in the commoditized banana market that is almost wholly controlled by multinationals mainly based in the USA.[107] Where globalization has resulted in the exploitation of peasants and labour, there are calls for the introduction of fair trade regimes from some quarters

---

[105]See R.B. Potter, T. Binns, J.A. Elliot and D. Smith, *Geographies of Development*, Harlow: Pearson Education, 2004.

[106]See B.H.Z. Moyo, "The Use and Role of Indigenous Knowledge in Small-scale Agricultural Systems in Africa: the Case of Farmers in Northern Malawi," Published PhD thesis. Glasgow University, 2008. http://theses.gla.ac.uk

[107]Press TV., Documentary. "Banana Cultivation in the Caribbean", 2010.

and deliberate considerations for sustainable agriculture.[108] Sustainable agriculture presents an opportunity to rethink the importance of family farms and rural communities.[109]

Intensification of production, therefore, is expressed mainly in form of uniform cultivars being grown as in Cavendish type of bananas and the dependence on employed agricultural labour in South America, for example. Furthermore, intensification refers to the requirement of having additional inputs such as mechanization to supplement hired labour in an effort to increase 'efficiency' and to maintain very high levels of output. However, the use of external inputs creates dependency on agro-industries including developments in biotechnology. In Malawi, tobacco industry has its own research unit responsible for both production of technologies and its dissemination to farmers, for example.

The consumers demand for uniform and quality products requires that particular skills are developed so that specialization both in skills and products become a necessity. A farm may as a consequence just specialize in one crop as opposed to mixed or intercropping. The principle of comparative advantage plays a major role in the process of specialization. Countries, regions and multinational companies specialize in agricultural products over which they have comparative advantage. Malawi has largely specialized in tobacco production, where it has lower production costs (advantages) over other producers. The principle of economies of scale reinforces the process of specialization leading to producing single crops on regional or farm level under mono-cropping as depicted by sugar cane growing in Malawi by Illovo Sugar Company. However, the need for diversification to reduce negative effects of price fluctuations and the need to reduce waste products encourages the processing of by-products that generate additional profits. Ethanol production associated with sugar production is a good example in Brazil and Malawi among many countries, for instance.[110]

---

[108] See University of California, "What is Sustainable Agriculture?" http://www.sarep.ucdavis.edu, 2010.

[109] Ibid.

[110] See D. Liwimbi, "Ethanol – the Spirit of Success," http://www.unep.org/urban_environment.

The impacts and understanding of globalization is highly skewed towards benefits that in reality accrue to developed countries. The information and communication technologies (ICT) have been largely expressed as the significant players in creating a single global village. Money, information and communication can move at a press of a button across the globe in a blink of an eye making the world nothing but one village. The benefits of ICT are indeed many in facilitating business across the globe. However, as the mobile phone, for example, benefits the fish and small-scale producer to transfer money and also find best markets for their products it does benefit most the technology provider. There remains a widening gap in wealth between the fish seller and the small-scale farmer on the one hand and the technology provider. Globalization remains a 'new' (another) tool for exploitation of developing countries up until fairness is re-invented.

3.3 Summary

Globalization is certainly not a new phenomenon. The speed of communication, money and technology transfer has significantly improved to very short time as in a touch of a button. The basic principle of having one global village has been recognized since people started travelling crossing oceans to other continents. The benefits of globalization are now threatening the very survival of some communities which are displaced both from their land and occupation and become proletariats only given the choice to work under multinational companies. Even governments lose their sovereignty in this process. The power of multinational companies and their wealth that eclipse some developing countries enable them to exploit resources and destroy the environment without concern.

Chapter 4

THE PURSUIT OF FOOD SECURITY BY SMALLHOLDER FARMERS IN MALAWI: THE CASE OF FARMERS IN ZOMBWE EXTENSION PLANNING AREA (EPA)

4.1 Introduction

In this chapter, the agricultural management practices at farm level are critically analyzed in terms of crops grown, farm size, number of gardens cultivated, the cultivation of wetlands (*dimba*) and the age of the household head. There is an attempt made to understand the use of indigenous knowledge by farmers to generate new knowledge, either by deliberate 'experimentation' or through simple curiosity.

4.2 Agricultural Management Practices at the Farm Level

Table 4.1 shows the diversity of the households in the study area in terms of household size, landholdings, age and number of gardens and *dimba* cultivated. Of paramount importance for this section are the numbers of gardens per household, the number of *dimba* cultivated and the mean farm size. These factors impact on how farmers allocate resources such as labour at any given time in their fields, and agricultural management practices at the farm level generally may vary because of the influences of these factors. These factors also show why diversifying agricultural production is a natural phenomenon under smallholder farming systems. Diversification enables farmers to enjoy diversity in operations and work loads that allows one to relax and enjoy agricultural activities 'normally' seen to be cumbersome, difficult and tiresome.

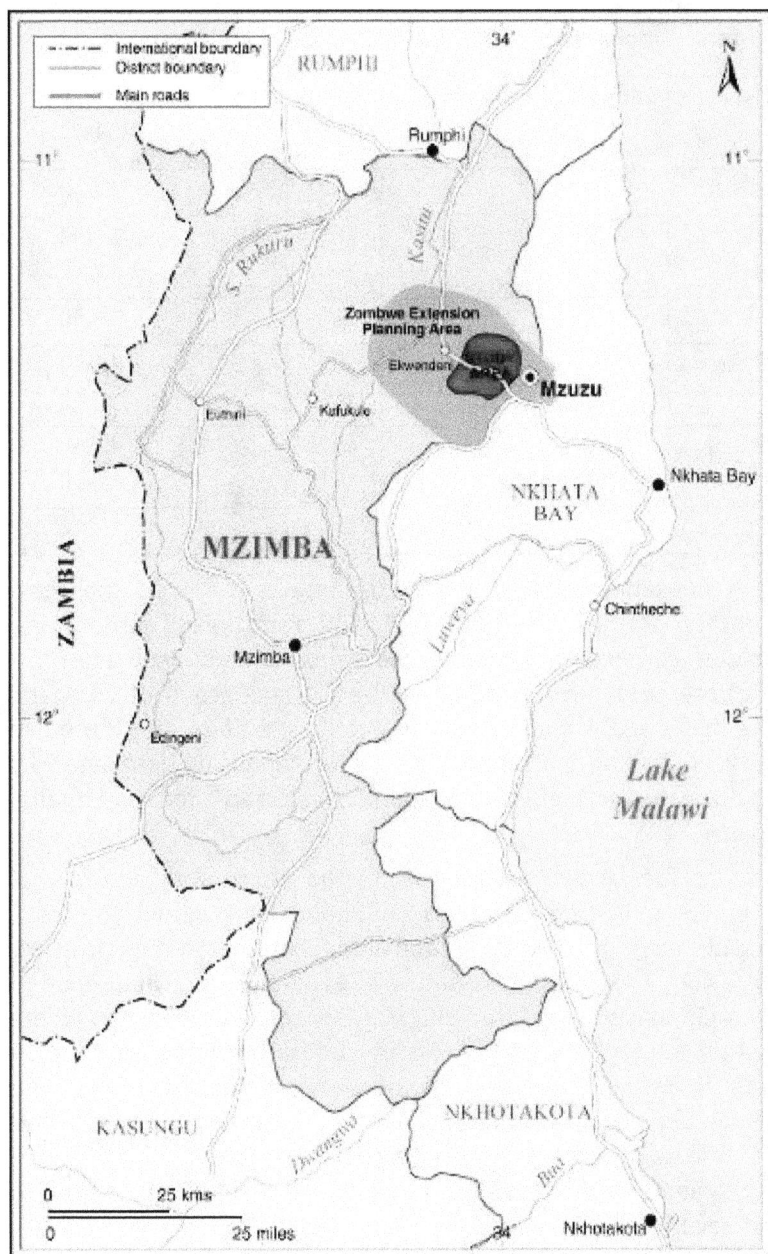

Figure 4.1 Zombwe Extension Planning Area

Table 4.1 Information on household characteristics of the study area

Characteristic	Minimum	Maximum	Mean	Mode	Standard Deviation
Household size	1	12	4.7	5	2.520
Gardens per household	1	6	2.9	3	1.037
Dimba	0	3	1.0	1	0.436
Cultivated area in hectares	1	20	3.9	5	2.030
Age of head of household	13	83	45.5	26	16.9
N=111					

The mean household size of the study area is 4.7 with the most frequently observed family size (mode) of 5, cultivating upland gardens that can be as big as 20 hectares excluding the *dimba* and as many as 6. The population of the study area is dominated by an age group of 26 years, although the mean age is 45.5 years. The modal age of 26 and the mean age of 45.5 years are an indication that the population is dominated by young people who are a labour source necessary for agricultural production.

The calendar for farming operations in the study area shows that operations in the upland gardens are continuous throughout the year, particularly with regard to harvesting food crops that is carried out in every month (Figure 4.2). This continuous nature of agricultural activities is often ignored when emphasis is towards rainy and dry season's activities where the rainy season is seen to be when most agricultural activities are done. Of course there is nothing wrong with this except that it places less importance on dry season operations that have an input in activities necessary in the rainy season. Harvesting is done throughout the year because of the amount of fruit and banana clones which farmers grow to ensure food security at household level and which mature almost every month and thus need to be harvested. The 'xx' on the calendar shows the overlap of major harvesting periods, which occur when cassava and maize are harvested at the same period of time during the months of April to Sep-

Figure 4.2 Calendar of farming operations in the study area in a year

Activities in upland gardens (*khonde* and *mundaukulu*)												
Activity	January	February	March	April	May	June	July	August	September	October	November	December
Land clearing	x						x	x	x	x	x	x
Ridging	x								x	x	x	x
Planting	x										x	x
Weeding	x	x	x									
Harvesting (fruits and crops)	x	x	x	xx	xx	xx	xx	xx	xx	x	x	x
Activities in dimba												
Land clearing				x	x	x	x	x				
Land tilling				x	x	x	x					
Planting				x	x	x	x	x				
Weeding				x	x	x	x					
Harvesting				x	x	x	x	x	x	x	x	x

tember. During these months between April and September maize in *mundaukulu* and *khonde* is mature while cassava is also considered as mature for a major harvest either for sale or processing into flour. Cassava is harvested throughout the year as a snack, while its leaves are eaten as a green vegetable boiled and taken with *sima*. There is very little waste from this crop. However, the main harvesting period for cassava begins in the month of April, which coincides with the harvesting of green maize. The continuous harvesting of cassava as a snack is not only useful as food in its own right, but is also a way of tasting and testing when cassava is suitable for major harvesting for selling to customers or for the preparation of flour suitable for making *sima*. From the months of April to September, cassava has a low water content, as compared to the rest of the year, and hence is more suitable for trading and processing into flour. Farmers monitor the properties of the cassava throughout the year by continuously tasting the crop. For farmers, knowledge has to be verified in agricultural management practices, such as the crop management, in order for them to make informed decisions. From experience, farmers are aware that variations in rainfall events can shift the suitable harvesting time from the beginning of the month of April in a 'drier' year to late April in a 'very wet' year. Farmers know about seasonal rainfall differences and their impacts on crop physiology, so that their management practices are influenced accordingly by the resultant conditions of their crops. The calendar double 'xx' thus shows increased levels of harvesting activities of these two crops (maize and cassava) alongside fruit harvesting.

Unlike along the shore of Lake Malawi, where cassava is still planted on *matutu,* is a staple food and thus harvested throughout the year for flour production (the factors of water content and suitability for good flour making becomes irrelevant), in the Mzuzu area, both cassava and maize are used in the preparation of *sima,* giving farmers a chance to diversify their diet according to their preferences for *sima* made from either maize or cassava. Although farmers in rural areas, especially in developing countries, have been criticized for being ignorant and backward, with their local knowledge labelled as static by some experts, here is an example of the importance of local knowledge in expressing and representing farmers' choices that make their lives fuller.[111] It becomes unhelpful to portray smallholder farmers in rural areas as 'struggling' to make a living when they innovatively and ingeniously utilize available resources as demonstrated by the use of cassava and maize to diversify their food sources.[112] It is important to note that most of the cassava grown along the shores of Lake Malawi for *sima* preparation is a bitter variety which cannot be eaten as a snack because it contains high levels of cyanide (personal communication with Bunda College of Agriculture nutritionists). The variety is deliberately chosen because of its high poison content (bitterness) and is reserved for processing into flour for *sima* production throughout the year. Its bitterness is a deterrent to those who would want to eat it as a snack, thereby reducing food security at the household level. The bitterness is diluted and dissolved by soaking cassava in water for 1-2 weeks, after which it is dried in the sunshine and pounded into flour having lost its bitterness and toxicity.

---

[111]See R. Chambers, *Rural Development: Putting the Last First.* London: Longman, 1983; R. Chambers, *Challenging the Professions: Frontiers for Rural Development.* London: Intermediate Technology Publications, 1993.

[112]See G.L. Beckford, Caribbean Peasantry in the Confines of the Plantation Mode of Production, *International Social Science Journal* XXXVII (3), 1985, pp 401-414.

In the study area, because the major staple food is maize, the cassava that is grown is sweet and is eaten as a snack, as well as being used for flour processing. There is deliberate selection of cassava varieties based on taste. It is interesting to note the importance of indigenous knowledge essentially to differentiate a single crop (cassava), that has many cultivars, based on just one common factor which is taste (bitter or sweet) in different places that are far apart (30 km or more). The knowledge about the taste of cassava can be transferred to areas beyond its production site. It is knowledge that can be utilized in the management of cassava to ensure food security at the household level throughout the whole of Malawi and possibly throughout the whole of Africa where cassava is produced.

Upland gardens are prepared from the months of July through to January (although occasionally some farmers might extend this period for reasons such as sickness, for example). This is a deliberate practice by farmers so as to spread the work effort over an extended period. There are two reasons for this. One is making the farmers' work easier to manage, and ensuring that other non-agricultural activities can be fitted in; and the other is that they then can plant crops such as maize over this extended time period so that they have a longer period in a year in which they have access to green maize, for example. The farming calendar reveals the strategic nature of indigenous knowledge use by the farmers. Farmers are able to extend the availability of green maize that they like to eat because of its sweetness and also occasionally sell for cash.

Ridging begins in the month of September and lasts up to the month of January. Between the months of September and October, ridges are not completely made, but only half made, which in Tumbuka is called *kuchelenga.* One side of the ridge is made, or just a small ridge of crop residues and weeds are made (*kukocheka).* This is done by placing weeds and crop residues using a hoe into an old furrow that will eventually be buried in the process of making a ridge. Ridges are made by digging soil from old ridges thereby burying weeds and crop residues placed in the furrow, a process technically known as banking. Banking is done by some farmers at a time when planting time is deemed imminent, necessary and convenient, based on past experience. The time at which ridging is done by farmers is influenced by their past experiences. It is important to note that at the time *kuchelenga* is done some farmers may not have decided when to plant, but know what is going to be planted. What to plant is a decision made several months before land preparation, and each garden has a major crop that dominates the mixed stand. It is still the dry season in the months of September and October and so other factors may determine when to plant, including the actual onset of the rainy season, yet some farmers have already started preparing their gardens. Some farmers plan and execute farming operations in advance of planting time, especially *kuchelenga* and the making of half ridges, both of which are viewed to be advantageous because subsequent ridging then uses less effort and is therefore completed quickly. As noted above, weeds and crop residues are buried under the soil when ridges are being made so that placing weeds and crop residues in the furrow in advance of ridging removes a stage that can delay the process. In addition, placing weeds and crop residues in furrows clears the farms of standing plants, making operations like tilling and ridging easier to accomplish. It also reduces the weed infestation in the gardens, as the completion of the ridge close to the day of planting kills the weeds that emerge with the light rains that occur throughout the year in the study area.[113]

Planting of maize, beans and other crops in the mixed stands begins in earnest in November after the 'planting' rains which generally occur towards the end of the month. Cassava is planted as a minor crop in fields dominated by other crops, particularly maize. When cassava is a minor crop, it is mainly consumed as the farmers visit their fields. To eat cassava, it has to be uprooted. Farmers have developed knowledge to plant the uprooted cassava stems immediately after uprooting them in spaces between the dominant crops in the mixed stand. Since the study area receives rain throughout the year, the planting of cassava is a continuous process, all year round. The stems are broken into 30 cm planting materials and inserted in the ridges of the main crop. The light rains throughout the year ensure its survival in the drier months. When ridging is conducted for new crops, the space occupied by cassava is left as a small mound, which in the second year forms part of the new ridge (Figure 4.3).

[113]See B.H.Z. Moyo, "The Use and Role of Indigenous Knowledge in Small-scale Agricultural Systems in Africa: the Case of Farmers in Northern Malawi," Published PhD thesis. Glasgow University, 2008. http://theses.gla.ac.uk

Figure 4.3 Cassava in small mounds in old furrows after maize harvest

Farmers noted that although weeding is indicated in the calendar as a major operation from the month of January onwards, such representation of weeding as an activity only depicts those operations conducted on the annual crops that are planted in the rainy season, while excluding the fact that weeding is done throughout the year. For many farmers, weeding is also done during frequent visits made to inspect crops. Such visits which are made on a regular basis are a form of inspection which enables farmers to see the conditions of their crops. Farmers also regularly visit virgin and/or fallow lands they own. Farmers pull out weeds that are close to them instinctively during such visits, including ferns in fallow and virgin lands because ferns are known to be difficult to weed when cultivation is subsequently undertaken. Such casual operations including weeding in virgin and fallow lands are rarely recognized is vital in conventional agriculture where operations are directly linked to crop production and weeding is associated with immediate crop output not necessarily the elimination of known species that are considered difficult weeds from experience.

The activities in *dimba* begin in earnest in the dry season from the end of the month of April. However, because perennial crops already exist here, the calendar ignores light weeding in the form of cutting grass around such crops. This cutting of grass using a *panga* is done as and when farmers consider that the weeds are a threat to crop survival. The process consists of a continuous evaluation of the perceived competition between plants and weeds. Just like in upland gardens, farmers uproot weeds as they inspect their crops. Land tilling is a major activity after land preparation that involves cutting grass and burning it. Water channels are made to drain excess water and then the land is tilled by first using a big hoe for deep cultivation, and then a light hoe to break the soil into finer particles suitable as a bed for crop planting (Figure 4.4).

Figure 4.4 Tilled land and dug water channels in dimba to drain water

For farmers, deep tilling is seen to lower the water table, while the breaking of these large soil chunks at a later stage is seen to reduce moisture loss by forming a physical barrier to moisture loss that was encouraged during the first tilling. Farmers know that a layer of fine soil is an effective barrier to moisture loss from evaporation in the hot months that follow. These forms of tilling are strategically timed, both to drain the soil of excess moisture, and then to protect it against further loss from subsequent crop production. Fine soil on a tilled surface is also used in conventional Western farming techniques to reduce moisture loss, a process called harrowing accomplished using an implement called a harrow. Local knowledge in Zombwe EPA about moisture conservation resembles the practices of modern farming. This detailed water management regime demonstrates how deep the farmers' knowledge is about wetlands (*dimba*) and how their moisture contents can be managed. Furthermore, some farmers plant the banks of the water channels with sugarcane to strengthen and stabilize them so that they can last beyond a single rainy season (Figure 4.5). It is easier to do maintenance work in subsequent years once the water channels are constructed, rather than to dig new ones each dry season. Farmers' operations are well planned and based on informed choices.

Major planting in *dimba* begins in June, mainly with green vegetables, such as rape, which is followed by beans. Beans are planted so that the flowering avoids the lowest temperatures experienced in the months of June and July.[114] Farmers have observed that very cold nights can kill bean flowers, so effort is made to avoid the loss of yield. Maize is then planted as the beans mature as a 'relay crop', so that it can be harvested when it is dry enough in December, just before the rains that can result in its destruction either by flooding or by mould infestation. This type of succession planting is seen to have advantages in reducing competition between plants, but also from fixing nitrogen by bean plants in the soil for the maize crop, despite the fact that in the uplands beans and maize are planted in the same planting stations. Farmers' observations over the years have made them manage these two crops differently in *dimba* and uplands. This is an interesting point that shows the high level of evaluation of practices under indigenous knowledge by farmers. They are able to compare upland crop management practices with those in *dimba* and make changes that suit

---

[114] Ibid.

Figure 4.5 Sugarcane planted on banks of a water channel

these different environments. Indeed, in some cases such as these, indigenous knowledge is not static, but is dynamic.[115] The knowledge about crop performance in these different environments is utilized by farmers so that effort put in their agricultural activities is rewarding and food security is ensured. The knowledge about upland and *dimba* management practices is derived from past experiences that involve 'experimentation' and careful observations of the performance of crops. The farming practices are different in uplands farms to those in *dimba* based on the results of farmers' 'experimentation' and the corresponding observations.

Indeed, many farmers use the *dimba* to 'multiply' planting materials such as sweet potatoes and new varieties of crops, such as cassava that are introduced by friends, government officials, private companies, non-government organizations (NGOs) and relatives. Farmers have in-depth

---

[115]See R. Chambers, *Rural Development: Putting the Last First*. London: Longman, 1983.

knowledge of *dimba* environments. The moisture that is available throughout the year in the *dimba* ensures the survival of such plants. Consequently, the *dimba* become seedbeds for new varieties before being tried in the main gardens. As seeds are multiplied, the characteristics of these new crops are observed creating knowledge to enable farmers to manage them, confirming the fact that agricultural management practices of crops in the study area (and probably across the nation and beyond) are evidence based on 'experimentation' and continual observation.

The harvesting of crops in the *dimba* begins as soon as the green vegetables are ready in the month of June and is completed with the harvesting of maize in the month of December. The months of January, February and March are a low activity time for *dimba* because they can be inundated as a result of the rains. Farmers make efforts through agricultural management practices to avoid crop destruction from water inundation. Cultivation of *dimba* is done so that crops mature and are harvested at times that reduce the destructive environmental factors, such as inundation, pests (including moulds) and low temperatures, affecting them.

4.3 Ensuring Food Security

The main priority of farmers is to ensure food security, not only at the household level but at the community level as well, something which is determined by the expectations and obligations of the extended family system. There are variations between families, to the extent to which farmers produce beyond family consumption needs. In this study (12) farmers have always produced for the market in addition to meeting family requirements. 'Market-oriented' production is used to raise cash for the purchase of items that are not produced on the farm. In addition, a major driving force for surplus production for sale is the need to pay for their children's school fees, something mentioned by almost all farmers.

The desire to have adequate food for the family drives farmers to grow many food crops. Table 4.2 shows that 107 farmers consider food security as 'very important' and only 2 farmers rated food security as only 'important'. No farmer ranked food security any lower than this. The importance of food security is also confirmed by 87 farmers who rank increased yield as 'very important' and 22 rank it as 'important'. There is no ambiguity about either the importance of food or secure yields on the part of all farmers. This point about food security has been demonstrated in the calendar of farming practices of the study area by the number of months farmers engage in crop production. Farmers are engaged in agricultural management activities that ensure food security throughout the year. They plant, weed, inspect and harvest food crops throughout the year. Crop management practices have been developed so that food is available all year round by either storing crops to last up to the next harvest or, in some cases, by growing crops that mature all year round, such as bananas and cassava. The government policy on ensuring food security thus conforms to the expectations of the local people and the consequent economic success enforces that that food security at household and community levels including national level is almost a guarantee for economic growth, a fact that is ignored in some instances where emphasis is placed on buy-in food rather than production. Indeed, the major successful economies have always ensured self-sufficiency in food.[116]

Table 4.2 Information on smallholder farmers' objectives for producing and protecting crops from pests and diseases.

Objectives	Very impor-tant	Impor tant	Neutral	Not impor-tant	Very unimpor-tant	Respon-dents	Total scores	%
Food security	107	2	0	0	0	109	543	99.6
Increase yield	87	22	0	0	0	109	523	96.0
Higher profit	12	60	32	2	1	107	400	73.4
Blemish-free	1	6	3	74	23	107	209	38.3
Scores	Very impor-	Impor tant=	Neutral = 3	Not impor-	Very unimpor-			

---

[116]See the USA, the EU after World War II and Japan.

	tant = 5	4		tant = 2	tant = 1			

All farmers interviewed said that they produce more food than is needed for household requirements. An example of one household serves as an illustration. The farmer has over 200 banana clones, over 200 fruit trees and over a hectare of cassava and maize, all of which is more than enough for his household requirements. Consequently, the farmer sells the excess for cash which adds to his other income from repairing radios, televisions and building houses in the study area.

The small number of 12 respondents that are of the opinion that higher profits are 'very important' says something about the nature and culture behind subsistence farming in the area. All farmers priorities the production of enough food for household consumption first, then any surplus production is sold for cash. However, food for household consumption is not just for the producing household. Farmers share production with friends and relatives. Food security is delivered only when all members of the community, and especially close family members, have adequate food.[117] A farmer is expected to produce enough for his/her own household, but also a surplus for other members of the extended family, neighbours and 'friends' that may fail to produce enough themselves on their own farms.[118] The case study of one farmer, who has a garden next to a feeder road used by children going to school, is used here to demonstrate the social aspects of food production within the communities in the study area and elsewhere in Malawi. As friends, school children and other people pass by when the farmer works on his garden, the farmer talks to them and typically uproots cassava for them to eat. This is a key social aspect of food production that is not fully appreciated when objectives such as profit-making are promoted by external agencies including development experts. However, this goes some way to explain the farmers' ranking of higher profit below food security; friendship and social capital mean sharing freely what is produced at the farm level. Mtika while working in southern Malawi

---

[117]See B.H.Z. Moyo, "The Use and Role of Indigenous Knowledge in Small-scale Agricultural Systems in Africa: the Case of Farmers in Northern Malawi," Published PhD thesis. Glasgow University. 2008. http://theses.gla.ac.uk; Mtika, 2000)

[118]M.M. Mtika, "Social Cultural Relations in Economic Action: the Embeddedness of Food Security in Rural Malawi amidst the AIDS Epidemic," *Environment and Planning* A 32, 2000, pp 345-360.

in Balaka demonstrates how labour is utilized to ensure food security. Labour is used to acquire food from those farmers who have (food) surplus.[119] Food is shared by those who provide labour not based on volume of work performed by each individual but equally so that each worker has adequate for consumption. This practice is at odds with Eurocentric/economic practice that rewards people based on equal pay for equal workloads.

Very few farmers (3) grow specific cash crops such as tobacco, which emphasizes this point of ranking food security in its broadest sense higher than profit. For farmers in the study area, food has a social value beyond feeding the immediate family and friends. It ensures social bonding, and the value of food can be more than 'just food'. The role and use of local knowledges must be appraised with the realization that communities have different values and norms so that poverty needs to be defined as reflected by the culture of the people concerned.[120] Here, it has been demonstrated that the prioritization of food security by farmers in Zombwe EPA and elsewhere in Malawi is underpinned by cultural factors. Food security is a community responsibility which is defined by cultural expectations.[121] Production levels are determined by household needs, extended families that may experience food shortages and festivals such as weddings, which are deliberately celebrated during the dry season when there is plenty of food from rain-fed crops and the weather is suitable for outdoor activities. Food production in the Zombwe EPA is in line with observations made by Escobar that societies have developed ways of defining and treating poverty that accommodates visions of community, frugality and sufficiency.[122] Indeed, farmers in Zombwe EPA produce food to accommodate the visions of the community including self-sufficiency at the household level.

---

[119]Ibid.

[120]See Arturo Escobar, *Encountering Development: the Making and Unmaking of the Third World*. Princeton University Press, 1995.

[121]See M.M. Mtika, "Social Cultural Relations in Economic Action: the Embeddedness of Food Security in Rural Malawi Amidst the AIDS Epidemic," *Environment and Planning* A 32, 2000, pp 345-360.

[122]Arturo Escobar, *Encountering Development: the Making and Unmaking of the Third World*. Princeton University Press, 1995.

The volatile prices of cash crops, such as tobacco, ensure the supremacy of food crops above cash crops in farmers' cropping strategies in some cases. For example, the prices of burley tobacco moved from an average of above one US dollar per kilogram to below 70 cents for a lengthy period in middle of the 1990s to 2006 and even lower in 2011.[123] Farmers are well aware of these price fluctuations and this discourages them from engaging in purely cash crop production and re-enforces the cultural practice of not depending much on the market. However, farmers in the study area did not raise the issue of prices of cash crops as an important factor in determining their priorities, even after probing them in relation to tobacco prices on the auction floors. The socially embedded value of food crops seems to eclipse the attraction of additional income from the production and sale of cash crops for many farmers in the study area. It also might be that the proximity of Mzuzu as an urban centre (market) creates alternative income sources from selling food crops, making them perform both as food as well as cash crops, an important factor for farmers because then they are able to achieve two of their objectives, food security and higher profit from the same crops. Indeed, all the farmers in this survey grow sugarcane as a food crop, as well as to provide cash when there is need, with only one farmer growing it for purely commercial purposes. His *dimba* is the only garden that has sugarcane as a dominant crop, occupying over 90% of the land area. Many farmers have sugarcane occupying only the banks of water channels with maize, beans and other crops, such as green vegetables, occupying the rest of the area in *dimba*.

While cash crops such as tobacco may have to be discarded if they become mouldy, for example, food crops such as beans that may be damaged by pests or diseases, can still be used, cooked as side dishes and eaten with *sima*. *Chilanda,* which is made from damaged beans, is prepared as a dish resembling mashed potatoes. Farmers, therefore, lose very little from blemished food crops because of their ability to use them, thus contributing to food security. Of course, if these were being produced as cash crops for the commercial standard market, they would be unsalable. However, such crops are sometimes sold amongst farmers for their own consumption; for example, one farmer said that he preferred to buy blemished beans because they are cheaper than blemish-free beans.

---

[123]See Mail and Guardian Online Newspaper, "Turmoil as Tobacco Prices Fluctuate in Malawi", 2008. Retrieved on 15th May 2008 from http://www.mg.co.za.

However, there is a small number of farmers who are committed to commercial-scale production, in addition to production for domestic consumption. There are 12 such farmers who rank higher profit as 'very important'. One exceptional case is the farmer producing broiler chickens for sale on a commercial scale. He produces 400 chickens which are sold every 6-8 weeks as broilers. He has developed traditional substitutes for commercial vitamin and mineral supplements, and treats chickens with his own locally-made medicines. He has also developed his own mixtures of vitamins A, B and C, and uses herbal medicines to prevent the common chicken disease in the area, Newcastle, from infecting his flock. There are efforts made by some farmers to move away from production only for home consumption to diversification for the market. Mzuzu provides them with an opportunity to do this, as it has a ready market for their produce. Their local knowledge is utilized for the production oriented to the market. Some farmers (10) produce maize at a commercial scale for sale, and they specifically grow hybrid maize for sale, because of its high 'bulky' yield, but grow local maize for home consumption. Local knowledge is dynamic and fits several situations in fulfilling farmers' diverse objectives. In the first instance, it ignored market forces in favour of local values and norms in the production of food crops; and in this latter case it has been fully used to benefit from the market opportunities offered by Mzuzu as a market outlet for the production of broilers, sugarcane and hybrid commercial maize.

Despite the fact that very few farmers priorities production for the market, farmers consider that their farming is sufficiently secure and productive that only a few food items that they cannot produce themselves, such as bread, butter and sugar, are sourced off the farm. One farmer responded *'usambazi nkulya'*, which means to be wealthy, one has to have food. It becomes safe to conclude that, although there are economic reasons for engaging in farming, farmers' understandings of production are based on being able to produce enough for domestic household consumption. It is food that is considered to be the wealth of the community and of the household. This is not surprising as, by their very nature, they are subsistence farmers, and most are only engaged in commercial production as an extension, or subsidiary activity, to subsistence farming.

It may not always be true, therefore, to assume that the lack of markets limits the extent to which small-scale farmers engage in commercial-scale production because this study shows that farmers produce mainly for home consumption as their overwhelming priority. For example, the growth of Mzuzu has increased the potential for commercial production, particularly for broilers and eggs. Yet, it is clear from this study that the presence of Mzuzu as a market outlet for farmers' produce has not significantly influenced many farmers to produce for it. It appears that sometimes, even where there are markets for products, not many farmers respond to the demand. Under subsistence farming systems, markets are typically regarded as places to dispose of food surpluses and to make sales that solve pressing cash needs, such as the payment of school fees, for example, and not necessarily to make profit as a major driving goal. The behaviour of local farmers in response to market forces is a challenge for development experts that they need to understand as they engage local people in development programmes that are based on modernization and market-led economic progress.

Many farmers in Zombwe Extension Planning Area (Figure 5.1) pointed out that they sold crops such as sugarcane to purchase household items (for example, clothes) and other foodstuffs that they do not produce. Others noted that selling of some crops including sugarcane was a measure to dispose of surpluses that would otherwise go to waste, as is the case, if sugarcane is left unused by end of the month of November. By this time, sugarcane is no longer suitable for consumption, as the canes become virtually tasteless because of the increased water uptake from *chizimyalupsya* rains that fall in October and/or November. Furthermore, if there was to be an increase in the production of sugarcane for commercial purposes, this would reduce the area put to other food crops, such as the staple maize, in *dimba,* something which is considered by most farmers as undesirable in the drive for food security.

The importance of food security is also demonstrated by the fact that farmers have established from their long experience in farming that *kusosa* and ridging are the only operations that can be spread over a long period of time, as critical timing is unnecessary and does not threaten crop output, as long as it is done in advance of the planting time. Planting and weeding by necessity cannot be spread over long periods of time, as the output of crops is negatively affected if and when such operations are delayed. Farmers are aware that weeds compete with crops for plant food in the soil and sunlight. They have noted that crops which are not weeded turn yellow and develop thin stems particularly if they are shaded. For farmers, weeds and shade increases the chance of reduced crop yields. Farmers make sure that an activity such as weeding is completed before damage to crop yields arises.

The importance of food security to farmers is further demonstrated by their planting at varying times, despite the fact that it has been demonstrated by agricultural extension staff, through field demonstration plots in the study area, that crops planted with the first 'planting' rains produce higher yields than those planted later. Thus, it would be realistic to expect higher yields from early planted crops, which then translates into increased food security at the household level. Utilizing the full rainy season improves the chances of increased yields. This is a factor particularly emphasized in recent years by agricultural extension staff, as a precautionary measure following three droughts in the last fifteen years in Malawi. The droughts have resulted in food deficits at both national and household levels, although the study area has not experienced food shortages. Extension workers apply blanket recommendations across different ecological regions. Despite this, farmers in the study area plant at times that differ from those advised by agricultural extension staff, which suggests that, although knowledge can be acquired through extension services by word of mouth or demonstrations, its use varies from farmer to farmer depending on circumstances faced by individuals.

It is important to note that the first 'planting' rains fall at different times in different years, and, as such, there is no definite date attached. It is clear that, despite the likelihood of yield loss that occurs as a result of planting maize after the first rains, farmers' considerations of food security go beyond yield potential, and become a matter of what actually is harvested and stored. Farmers' understandings of food security are based on what eventually gets into their storage system, and it is this that determines the nature of their farming practices. The potential yield, as demonstrated under ideal conditions of demonstration plots managed by extension workers, is of little interest to farmers, especially if, by trying to achieve the potential yields, it means increasing the danger of facing an even lower output than is normally achieved using proven methods of production, such as planting after the first rains, especially around Christmas time. Farmers' management practices, including the application of fertilizer, crop varieties grown and weeding, are different from those carried out on demonstration plots. Demonstration plots are planted with hybrid varieties, kept weed free and receive recommended rates of fertilizers (4-8 bags per hectare) which most farmers cannot afford, thereby putting their chances of realizing the potential yield of hybrids out of reach. Hybrid crops, especially maize, are known to farmers to be susceptible to pests particularly those which attack the crop in storage, such as weevils. The destruction of crops in storage increases the chance of reducing food security at the household level. The knowledge which farmers have about demonstration plots in relation to potential yields is based on past experiences. It is safe to conclude that farmers' knowledge is grounded and based on past performances and results. When the past performance has resulted in high yields, it then becomes useful knowledge, which is retained for as long as it remains useful and productive. At such a point, Western technologies fail to displace it.

There is another reason advanced by farmers for planting maize at different times at the beginning of the rainy season. Many farmers explained that maize planted earlier than the Christmas period tends to be heavily infested with maize stalk-borer, and therefore the yields gained by early planting are reduced by stalk borer infestation. Indeed, transect walks by the researcher established that up to 40% of the crop planted earlier than the Christmas period showed signs of severe (50% of a leaf area or 60% of a stem affected) stalk-borer infestation (Figure 5.6). Farmers' knowledge is based on what they see and this is evaluated for its impact on yield. Stalk-borer infestation in maize is understood to reduce yields by as much as 30%, because a highly infested plant can fail to bear a cob. This is a significant point which farmers cannot ignore in the process of ensuring food security.

Farmers understand that, in order to be food secure, their farming practices need to protect crops from those diseases and pests that attack them either in the field or in storage. Farmers have a number of options, one of which is to apply chemical pesticides to protect crops. For many, it is easier and certainly cheaper to avoid pests in the fields by varying planting times, as demonstrated by their varying times of planting maize to escape its major pest (maize stalk-borer). Therefore, most farmers (59%) prefer to plant their main gardens around Christmas time, as past experience has shown that planting then generates minimum risk of stalk-borer infestation in their main staple crop. In high value crops, like tomatoes, pests and diseases are treated with the application of pesticides.

Farmers in the study area cultivate a fairly large area (3.9 hectares) given that the labour force is raised from the family whose average size is 5. To manage such an area effectively (3.9 hectares) demands the spread of

Figure 4.6 Evidence of maize stalk borer infestations in an early planted upland garden.

some activities over a longer period of time. Planting is therefore spread from the month of November to the month of January in response to low labour availability at the household level, although, as noted earlier, this has the advantage of increasing the length of time farmers have green maize available for consumption. The additional advantage is that this creates an opportunity for farmers to see the extent to which crops are attacked by pests and diseases with respect to different planting times. This is in line with observations made by many scholars engaged in studying indigenous knowledge that its production is dynamic and based on continuous 'experimentation' and observation (for example, Chambers, 1983). Indeed, the observations made become useful knowledge as to when it is best to plant crops in order to escape diseases and pests that attack them on a yearly basis. The knowledge generated, therefore, is used in a realistic and practical way to achieve food security in their farming practices. The relatively large area under cultivation (3.9 hectares) in the study area has an additional aspect to it; it is viewed by farmers as essential because it increases the chance of having enough food for household consumption. An old adage that 'size matters' comes into play in the case of cultivated areas. For farmers, although some crops may be damaged by infection and pest infestation that arise from the different planting times, enough is seen to be available as a result of their fairly large cultivated areas. If their farms were small, the risk of crop damage by pests and diseases would be seen to be too high to take and planting would be limited to proven safe periods, such as only after Christmas. Indeed, farmers with garden sizes around one hectare plant during this safe period, although it might be a factor related to the amount of labour available, as pointed out above. The landholding size and labour availability have implications for farming practices that have both negative and positive aspects, as demonstrated here. Landholding size is a major factor in influencing farming practices, especially as there is no mechanization to accelerate the completion of activities.

There is another important geographical factor that influences the time of planting crops such as maize in upland gardens. Farmers' most fertile sites are ant hills. By their nature of a high clay content and steep slopes, the soil is not sufficiently moist after the first 'planting' rains. Rainwater runs off, limiting percolation. The ant hills' characteristics of delayed water uptake and resultant late soaking lead many farmers to plant only when they have acquired adequate moisture for seed emergence and this is normally only achieved after several days (7-10 days) of rain above 25 mm. Indeed, farmers said that *chidulichi wombe dankha ndi poti kugoma*. This translates as the ant hill has to be adequately moist before planting. They know that planting before such rains results in the seed failing to emerge because of dry soil on its base, or the top surface being too hard and sealed off so that the shoot fails to break through the surface layer, especially if there is a break in the rain at the time of shoot emergence. Farmers noted that digging the planting stations have shown twisted shoots in the ground. This is a proven and tested outcome of ant hill management practice that cannot be ignored, so that farmers make strenuous efforts to ensure that planting crops is done according to this well-established knowledge repertoire. The drive to ensure food security by farmers in the study area has major impacts and implications for their agricultural management practices.

4.4 Fundamentals of African Agriculture

As we have seen, farmers in Malawi and many traditional ways of cultivation, the major aim is to maximize the chance of ensuring food security for sustenance of life. However, many scholars have looked at the farmers' practices as risk spreading.[124] This position of scholars is based on Western view point which emphasises on principles of gambling.[125] Gambling requires reducing risk of losing and involves calculation of the probability of winning. This is then uncritically applied to African ways of farming as being risk averse or more especially where there is limited knowledge of some scholars in the African ways of knowing.[126]

It is noted here in this book and elsewhere that farmers practice mixed cropping or intercropping, agro-forestry, keep livestock in addition to growing crops and even spreading the livestock owned by letting some of it to be kept by friends and relatives. This practice may be better understood as increasing the opportunities for success than a function of risk spreading activity. The farmers understand that success arises from diversity of activities. The same advantages associated with increasing and maintaining diversity in the ecosystem for its resilience is replicated by farmers in their operations. The farmers' actions associated with planting various crops and rearing livestock is based on positivity. They plant crops at a time when

---

[124]See B. Legesse, *The Need for an Integrated Theoretical Framework for Understanding Smallholder Farmers' Risk Perceptions and Risk Responses: Review Essay and some Experiences from Ethiopia*, University of Uppsala, Sweden, 2002.

[125]See N. Han, J.F. Shogren and B. White, *Introduction to Environmental Economics.* Oxford University Press, 2001, pp 123; Lanning and Mueller, *Africa Undermined: Mining Companies and the Underdevelopment of Africa*, Harmondsworth, Penguin, 1979.

[126]See C. Eriksen, "Why do they Burn the 'Bush'? Fire, Rural Livelihoods, and Conservation in Zambia," *The Geographical Journal* 173 (3), 2007, pp. 242-256; T.C. Phuthego and R. Chanda, "Traditional Ecological Knowledge and Community Based Natural Resources Management: Lesson from Botswana Wildlife Management Area," *Applied Geography* 24 (1), 2004, pp. 57-76.

they are sure they will succeed based on careful experimentation and consequent observations. They also keep various types of livestock (chicken, cattle, pigeons, pigs, sheep, guinea fowls etc) as a means to increase the chance of success – utilizing opportunities provided for in diversity and not necessarily trying to minimize risks. As in most cases the results and outcome of planting time is at the mercy of the unpredictability of Mother Nature so too is the survival chances of livestock. However, such problems associated with nature including drought are compensated for by dimba cultivation, planting of drought tolerant and resistant crops and keeping of livestock that can enable farmers survive through droughts, crop and livestock diseases and pests and climate change. In a way farmers retain and apply the principles of comparative advantage by growing crops and rearing of livestock best suited to their areas. The economies of scale are realized from the increased number of crops grown and livestock kept including increased number of gardens cultivated. That the market is a tool for use when there is need is clear from the diversity of their practices and enterprises. They produce almost all what they need. For farmers in Africa there is no need to specialize and be dependent on the market when it is imperfect. The 2007-8 credit crunch and the highly indebted developed countries (Ireland, Greece, Italy, Portugal and Spain) position today (2011) confirms the unreliability of the market in providing adequately for the population. The unreliability of the market is further illustrated in trade barriers and quota systems used to protect farmers in developed countries from competition from the efficient agricultural producers under smallholdings in developing countries.[127] The farmers in Malawi and Africa in general have been aware of the imperfect nature of the market for generations and have by design developed production systems that are less dependent on the market.

4.5 Summary

It is clear from this chapter that smallholder farmers have a wide knowledge concerning their operations and these vary depending on the environment, type of crops and livestock, climate, weather variability and experience. What is significant here is that one has to realise that

---

[127]See M. Redclift, *Sustainable Development: Exploring the Contradictions*, London and New York: Routledge, 1995.

smallholder farmers across the country and beyond have detailed knowledge in what they are engaged in. The nature of the knowledge may be different but it has evolved to enable them leave a fuller life in their environment. It is unfortunate that at times these farmers embark on unsustainable practices because of forces beyond their control when they have the knowledge to manage resources properly even for the benefit of future generations.

References

Adriansen, H.K., "Understanding Pastoral Mobility: the Case of Senegalese Fulani," *The Geographical Journal* 174 (3), 2008, pp. 207-222.

Bebbington, A., "Modernization from Below: an Alternative Indigenous Development?" *Economic Geography* 69, 1993, pp. 274-292;

Beckford, C., D. Barker and S. Bailey, "Adaptation, Innovation and Domestic Food Production in Jamaica: Some Examples of Survival Strategies of Small Scale Farmers," Singapore Journal of Tropical Geography 28, 2007, pp 273-286.

Beckford, G.L., Caribbean Peasantry in the Confines of the Plantation Mode of Production, International Social Science Journal XXXVII (3), 1985, pp 401-414.

Black, J., 'O ye Green Memories O' The Auld Days.' A Renfrewshire Farmer's Story. Dundee: Accolade Publishing, 2006.

Blaikie, P., "Development, Post-, Anti-, and Populist: a Critical Review," *Environment and Planning* A 32, 2000, pp. 1033-1050;

Brady, N.C., "The Nature and Properties of Soils". New Jersey, Prentice Hall. 1974.

Briggs, J. "The Use of Indigenous Knowledge in Development: Problems and Challenges," *Progress in Development Studies* 5 (2), 2005, pp. 99-114;

Briggs, J. and J. Sharp, "Indigenous Knowledges and Development: a Postcolonial Caution," Third World Quarterly 25 (4), 2004, pp. 661-676.

Briggs, J., J. Sharp, H. Yacoub, N. Hamed and A. Roe, "Environmental Knowledge Production: Evidence from the Bedouin Communities in Southern Egypt." Egypt Journal of International Development 19, 2007, pp. 239-251;

Briggs, J., M. Badri and A.M. Mekki, "Indigenous Knowledges and Vegetation Use among Bedouin in the Eastern Desert of Egypt," *Applied Geography* 19, 1999, pp. 78-103.

Brohman, J., *Popular Development: Rethinking the Theory and Practice of Development*. Oxford, Blackwell, 1996.

Buchanan, J., *The Shire Highlands (East and Central Africa) as Colony and Mission*. Edinburgh and London, William Blackwood and Sons, 1885.

Chambers, Robert, *Challenging the Professions: Frontiers for Rural Development*. London: Intermediate Technology Publications, 1993.

Chambers, Robert., *Rural Development: Putting the Last First*. London: Longman, 1983.

Chienga, C., Effects of Variety and Processing Methods on Physical and Chemical Characteristics of Maize Flour and Sensory Properties of Nsima, MSc, University of Malawi, Bunda College of Agriculture, Lilongwe, 2012.

Chirwa, E.W., "Access to Land, Growth and Poverty Reduction in Malawi". University of Malawi, Chancellor College, Zomba, 2004.

Chirwa, E.W., "Access to Land, Growth and Poverty Reduction in Malawi Policy Brief". University of Malawi, Chancellor College, Zomba, 2005.

Eriksen, C., "Why do they Burn the 'Bush'? Fire, Rural Livelihoods, and Conservation in Zambia". *The Geographical Journal* 173 (3), 2007, pp. 242-256.

Escobar, Arturo, *Encountering Development: the Making and Unmaking of the Third World*. Princeton, Princeton University Press, 1995.

Forbes, D.K., *The Geography of Underdevelopment*. London, Croomhelm, 1986.

Gomes de Almeida, S., and Fernandes, G.B., "Economic Benefits of a Transition to Ecological Agriculture. Changing Farming practices". *LEISA Magazine* 22 (2), 2006, pp. 28-29.

Han, N., Shogren, J. F., and White, B., 2001. *Introduction to Environmental Economics.* Oxford University Press, pp 123.

Hansen, N., "Development from Above: The Centre-Down Development Paradigm," Editors W.B. Stohr, and D.R.F Taylor, *Development from Above or Below? The Dialects of Regional Planning in Developing Countries*, Chichester: John Wiley and Sons, 1981, pp. 15-38;

Harrison, G., "Peasants, the Agrarian Question and Lenses of Development," *Progress in Development Studies* 1 (3), 2001, pp. 187-203.

Hove, H., "Critiquing Sustainable Development: A Meaningful Way of Mediating the Development Impasse?" Undercurrent vol 1(1), 2004, pp 48-54. Retrieved on 6th June 2008 from http://forbin.mit.edu/Ris_and_Preferences/decision models.jsp.

Hudson, N., *Soil and Water Conservation*. Bedford: Silsoe Associates Ampthill. 1981.

Jones, J.G., "Pollution from Fish Farms" in *Agriculture and Environment*, Editor J.G. Jones, New York: Ellis Horwood, 1993.

Kratli, S., "What do Breeders Breed? On Pastoralists, Cattle and Unpredictability". *Journal of Agriculture and Environment for International Development* 102 (1-2), 2008, pp. 123-139.

Lado, C., "Sustainable Environmental Resource Utilisation: a Case of Farmers' Ethno Botanical Knowledge and Rural Change in Bungoma District, Kenya". *Applied Geography Journal* 24, 2004, pp. 281-302.

Langyintuo, A., Malawi Maize Sector Stakeholders' Workshop Report: Strengthening Seed Marketing Incentives in Southern Africa to Increase the Impact of Maize Breeding Research Project. International Maize and Wheat Improvement Centre (CIMMYT), Harare, Zimbabwe, 2004.

Lanning, G., and M. Mueller, 1979. Africa undermined: mining companies and the underdevelopment of Africa. Harmondsworth, Penguin Books, 1979.

Legesse, B., The Need for an Integrated Theoretical Framework for Understanding Smallholder Farmers' Risk Perceptions and Risk Responses: Review Essay and some Experiences from Ethiopia, University of Uppsala, Sweden, 2002.

Lele, U. "Structural Adjustment, Agricultural Development and the Poor, Lessons from the Malawian Experience": The International Bank for Reconstruction and Development/The World Bank, 1989. Retrieved on 28th October 2005 from http://www-wds.worldbank.org.

Leys, C., *The Rise and Fall of Development Theory*. London, James Currey, 1996.

Liwimbi, D., "Ethanol – the Spirit of Success". http://www.unep.org/urban _environment.

Mail and Guardian Online Newspaper, "Turmoil as Tobacco Prices Fluctuate in Malawi", 2008. Retrieved on 15th May 2008 from http://www.mg.co.za.

Malawi Nation [Newspaper], "Tobacco Business: who is Duping the Poor Farmer?" 2008. Retrieved on 15th May 2008 from http://www.nationmw.net.

Morse, S., "The Geography of Tyranny and Despair: Development Indicators and the Hypothesis of Genetic Inevitability of National Inequality". *The Geographical Journal* 174 (3), 2008, pp. 195-206.

Moyo, B.H.Z., "Indigenous Knowledge-Based Farming Practices: A Setting for the Contestation of Modernity, Development and Progress". *Scottish Geographical Journal* 125 (3-4), 2009, pp. 353-360.

Moyo, B.H.Z., "The Use and Role of Indigenous Knowledge in Small-scale Agricultural Systems in Africa: the Case of Farmers in Northern Malawi," Published PhD thesis. Glasgow University, 2008. http://theses.gla.ac.uk

Mtika, M.M., Social Cultural Relations in Economic Action: the Embeddedness of Food Security in Rural Malawi Amidst the AIDS Epidemic". *Environment and Planning* A 32, 2000, pp 345-360.

Munck, R., "Political Programmes and Development: The Transformative Potential of Social Democracy." Editor F.J. Schuurman, *Beyond the Impasse: New Directions in Development Theory*, London: Zed Books, 1993, pp 113-139.

Narman, A., "Getting Towards the Beginning of the End for Traditional Development Aid: Major Trends in Development Thinking and its Practical Application over the Last Fifty Years"; Editors D. Simon and A. Narman, *Development as Theory and Practice*, Harlow: Longman, 1999, pp. 149-180;

Peters, P.E., "The Limits of Knowledge: Securing Rural Livelihoods in a Situation of Resource Scarcity," Editors C.B. Barrett, F. Place and A.A. Aboud, *Natural Resources Management in African Agriculture Understanding and Improving Current Practices*, Oxon: CABI Publishing, 2002.

Phiri, D.D., "D.D. Phiri Column, Lessons from Green Revolution". *The Nation* [Newspaper] of 26th March, 2010.

Phuthego, T.C., and Chanda, R., "Traditional Ecological Knowledge and Community Based Natural Resources Management: Lesson from Botswana Wildlife Management Area". *Applied Geography* 24 (1), 2004, pp. 57-76.

Pieterse, J.N., "Critical Holism and the Tao of Development" Editors R. Munck and D. O'Hearn, *Critical Development Theory: Contribution to a New Paradigm*, London: Zed Books, 1999, pp. 63-88.

Potter, R.B., T. Binns, J.A. Elliot and D. Smith, *Geographies of Development*, Harlow: Pearson Education, 2004.

Power, M., *Rethinking Development Geographies*. London, Routledge, 2003.

Press TV., Documentary. "Banana Cultivation in the Caribbean", 2010.

Redclift, M., *"Sustainable Development: Exploring the Contradictions"*. London and New York: Routledge, 1995.

Riseth, J.A., "An Indigenous Perspective on National Parks and Sami Reindeer Management in Norway". *Geographical Research* 45 (2), 2007, pp. 177-185.

Sardar, Z., "Development and the Location of Eurocentrism," Editors R. Munck and D. O'Hearn, *Critical Development Theory: Contributions to a New Paradigm*, London: Zed Books, 1999;

Schuurman, F.J., "Development Theory in the 1990s," Editor F.J. Schuurman, *Beyond the Impasse: New Directions in Development Theory*," London: Zed Books, 1993

Simon, D., "Beyond Antidevelopment: Discourses, Convergence, Practices". *Singapore Journal of Tropical Geography* 28, 2007, pp. 205-218.

Stringer, C.S., C. Twyman and L. Gibbs, "Learning from the South: Common Challenges and Solutions for Small-Scale Farming," *The Geographical Journal* 174 (3), 2008, pp. 235-250.

Taiwo, L.B. and B.A. Oso, "Influence of Composting Technologies on Microbial Succession, Temperature and pH in a Composting Municipal Solid Waste. "*African Journal of Biotechnology*, 3(4) pp 239-243.

Tevera, D.S., "Do they Need Ivy in Africa? Ruminations of an African Geographer Trained Abroad," Editors D. Simon and A. Narmann, *Development as Theory and Practice*, Harlow: Longman, 1999.

Thompson, I.B., "The Role of Artisan Technology and Indigenous Knowledge Transfer in the Survival of a Classic Cultural Landscape: the Marais Salants of Guerande, Loire-Atlantique, France," *Journal of Historical Geography* 25 (2), 1999, pp. 216-234.

Townsend, J., "Gender Studies: Whose Agenda?" Editor F.J. Schuurman, *Beyond the Impasse: New Directions in Development Theory*, London: Zed Books, 1993.

Tucker, V., "The Myth of Development: a Critique of a Eurocentric Discourse," Editors R. Munck and D. O'Hearn, Critical Development Theory: Contribution to a New Paradigm," London: Zed Books, 1999, pp 1-26.

University of California, "What is Sustainable Agriculture?" http://www.sarep. ucdavis.edu, 2010.

World Bank., "Indigenous Knowledge for Development: a Framework for Action. Knowledge and Learning Centre", Africa Region, World Bank, 1998. Retrieved on 12th April 2007 from http://www.worldbank.org/afr/ik.

www.ingramcontent.com/pod-product-compliance
Lightning Source LLC
Chambersburg PA
CBHW052013270326
41929CB00015B/2905